Steganography Techniques for Digital Images

Abid Yahya

Steganography Techniques for Digital Images

Springer

Abid Yahya
Faculty of Engineering & Technology
Botswana International University of Science and Technology
Palapye, Botswana

ISBN 978-3-030-13263-7 ISBN 978-3-319-78597-4 (eBook)
https://doi.org/10.1007/978-3-319-78597-4

Softcover re-print of the Hardcover 1st edition 2019

Printed on acid-free paper

This Springer imprint is published by the registered company Springer International Publishing AG part of Springer Nature.
The registered company address is: Gewerbestrasse 11, 6330 Cham, Switzerland

Dedicated to my family for their love, support, and sacrifice along the path of my academic pursuits, especially to my father, who took me to school.

Abid Yahya

Preface

Transmitting data over a public network—for instance, the Internet—necessitates increasing the security of data communications, especially within the extremely sensitive category of file transfers. Steganography systems have been presented and established to offer security for these applications. Profoundly, the steganography objective is not only to obstruct the adversary from decrypting a hidden message, but also to divert an adversary from even suspecting the presence of concealed communications. It does not constitute cryptography; however, it improves security using its obscurity features. If an adversary's suspicion is aroused while examining a document where steganography has been employed, the goal of the latter objective will be overpowered, irrespective of whether or not any plaintext is discovered.

This book is comprised of five (5) chapters, and is prepared as follows:

Chapter 1 presents an overview and the challenges in the field of steganography that clarify the driving force and motivating aspect, in conjunction with the objectives, and a book outline.

Chapter 2 describes the background and history of digital-image steganography. The evaluation jumps by connecting the present work to other existing investigations in the literature. This is achieved by up-to-date high-tech improvement in the field of digital-image steganography. Assessments and analyses of each process are also delivered where possible. Furthermore, the chapter gives a taxonomy of the present steganographic systems on image files.

Chapter 3 gives a detailed explanation of the design of a new image steganography algorithm, known as CR-BIS. The proposed scheme takes advantage of the characteristic region-detection algorithm and content-based embedding technique. Moreover, the chapter discusses in detail the processes and stages of the algorithm and the benefits it brings over existing algorithms.

Chapter 4 inspects in detail the theoretical features of an improved robust and reliable steganographic system, by considering the advantages of CR-BIS algorithm along with applying the insertion of an embedding map. Moreover, the chapter demonstrates the ability to enhance the stego-system robustness either by ECC inclusion or by adding redundant bits to the secret message to be embedded.

Chapter 5 concludes the book by summarizing the most important ideas, conclusions, contributions, and future direction in the field of hidden communication.

Palapye, Botswana

Abid Yahya
Nagham Hamid
R. Badlishah Ahmad
Joseph M. Chuma

Acknowledgments

Thank you to the following personalities without their contributions and support, this book could not have been written.

The authors are interested in expressing their extraordinary appreciation and obligations to Botswana International University of Science and Technology (BIUST), Universiti Malaysia Perlis (UniMAP), Sarhad University of Science & Information Technology, Peshawar, Pakistan (SUIT) and Muni University, Arua, Uganda for giving us the consideration, time, and opportunity to publish this book.

We would also like to take this opportunity to express our gratitude to all those people who provided us with invaluable help in the writing and publication of this book.

Contents

About the Authors

Abid Yahya began his career on an engineering path, which is rare among other research executives, and he earned his MSc and PhD degrees in wireless and mobile systems from the Universiti Sains Malaysia, Malaysia. Currently, he is working at the Botswana International University of Science and Technology. He has applied a combination of practical and academic experience to a variety of consultancies for major corporations.

He has more than 115 research publications to his credit in numerous reputable journals, conference articles, and book chapters. He has received several awards and grants from various funding agencies and supervised a number of PhD and master's candidates. His new book, *LTE-A Cellular Networks: Multi-hop Relay for Coverage, Capacity and Performance Enhancement*, was published by Springer International Publishing in January 2017 and is being followed in national and international universities.

Professor Yahya was assigned to be an external and internal examiner for postgraduate students. He has been invited a number of times to be a speaker or visiting lecturer at various multinational companies. He sits on various panels with the government and other industry-related panels of study.

Nagham Hamid earned her BSc degree in electronic and communications engineering from Al-Nahrain University, Baghdad, Iraq. She got her MSc degree in Modern Communication from Al-Nahrain University as well. In 2013, she got her PhD degree from University Malaysia Perlis in Communication Engineering, majoring in Information Security. Currently she is a lecturer in Dijlah University College in Computer Engineering department, Baghdad, Iraq.

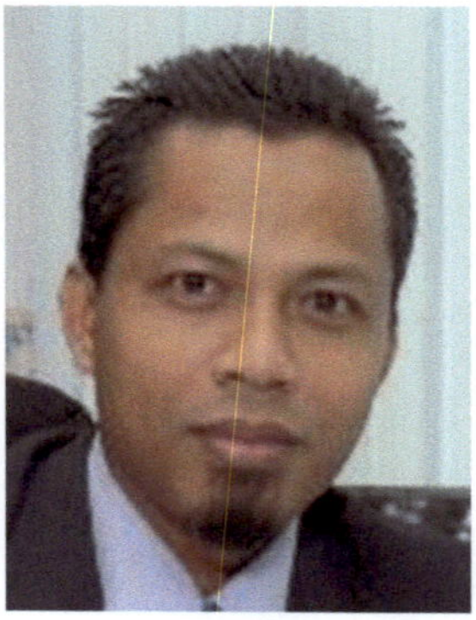

R. Badlishah Ahmad obtained his BEng in Electrical & Electronic Engineering from Glasgow University in 1994. He obtained his MSc and PhD in 1995 and 2000 respectively from the University of Strathclyde, UK. His research interests are on computer and telecommunication network modeling using discrete event simulators, optical networking & coding and embedded system based on GNU/Linux for vision. He has five (5) years teaching experience in Universiti Sains Malaysia. From 2004 until now, he has worked with Universiti Malaysia Perlis (UniMAP) as the Dean of the School of Computer and Communication Engineering.

Joseph M. Chuma received a BEng in Electrical and Electronic Engineering from the University of Nottingham, UK in 1992, an MSc in Telecommunications and Information Systems and a PhD in Electronic Systems Engineering from the University of Essex, UK in 1995 and 2001 respectively. His main areas of research are in the design of compact single- and dual-mode dielectric resonator filters for mobile communications. Prof Chuma has served as the Dean of the Faculty of Engineering and Technology at the University of Botswana. He is also serving as a Board Member in a member of Parastatals Organisation in Botswana. Currently, he is working as HOD in the Department of Electrical, Computer and Telecommunications Engineering at the Botswana International University of Science and Technology.

List of Abbreviations

ADR	Accuracy of Correctly Detected Characteristic Regions
AES	Advanced Encryption Standard
AET	Image Adaptive Energy Thresholding
A-MSPU	Adaptive More Surrounding Pixels Using
ANNTS	Artificial Neural Network Technology for Steganography
AWGN	Additive White Gaussian Noise
BER	Bit Error Rate
BCH	Bose-Chaudhuri-Hochquenghem
BMP	Bitmap Format
BSSIS	Blind Spread Spectrum Image Steganography
bpp	Bit per Pixel
CD	Compact Disc
CDF	Cohen-Daubechies-Feauveau
CDMA	Code Division Multiple Access
CPA	Chosen-Plaintext Attack
CR-BIS	Characteristic Region-Based Image Steganography
dB	Decibel
DCT	Discrete Cosine Transform
DES	Data Encryption Standard
DFT	Discrete Fourier Transforms
DoG	Difference-of-Gaussian
DS/FH	Direct Sequence/Frequency Hopping
DVD	Digital Video Disc
DWT	Discrete Wavelet Transforms
ECC	Error Correcting Coding
FCM	Fuzzy C-Means
GA	Genetic Algorithm
GF	Galois Field
GIF	Graphics Interchange Format
HC-RIOT	Homogenous Connected-Region Interested Ordered
HCSSD	High Capacity and High Security Steganography System

HVS	Human Visual System
ID	Identity Card
IRSS	Improved Robust And Secured Steganography
JPEG	Joint Photographic Experts Group
LDPC	Low-Density Parity-Check
LSB	Least Significant Bit
MB	Model-Based Steganography
MBNS	Multiple Base Notational Systems
MLE	Maximum Length Embeddable
MMS	Multimedia Messaging Service
MSE	Mean Squared Error
MVE	Mean Value of Energy
PNG	Portable Network Graphics
PSNR	Peak Signal to Noise Ratio
PRNG	Pseudo-Random Number Generator
PVD	Pixel-Value Differencing
QF	Quality Factor
RBGC	Reflected Binary Gray Code
RGB	Red Green and Blue
RS-Code	Reed–Solomon Code
SIFT	Scale-Invariant Feature Transform
SINR	Signal to Interference Plus Noise Ratio
SNR	Signal to Noise-Ratio
SS	Spread Spectrum
SSIS	Spread Spectrum Image Steganography
SURF	Speeded-Up Robust Features
TIFF	Tagged Image File Format
VBs	Valid Blocks
VCs	Valid DCT Coefficients
WMF	Windows Metafile
XOR	Exclusive-Or-Operation
YASS	Yet Another Steganographic System

Chapter 1
Introduction to Steganography

Abstract With the expansion in digital-communication technologies and the rapid growth of network bandwidth, the Internet has turned out to be a commonly used channel for transmitting many documents—for instance, audio, video, image, and text—in digital form. Many practices have been offered and developed for providing the secure transmission of data. The focus of the current research is on the design of data-hiding techniques used for transmitting secret data where digital images are selected as the cover-media. This chapter has identified the problems in the present image-steganography schemes.

1.1 Introduction

With the expansion in digital communication technologies and the rapid growth of network bandwidth, the Internet has turned out to be a much-used channel for transmitting many documents—for instance: audio, video, image, and text—in digital form. Many practices have been offered and developed for providing the secure transmission of data. A common approach to providing the secure environment for important data transmission is the use of cryptographic techniques (Ling, 2005).

Cryptography transforms data into seemingly meaningless bits, called cipher text, by using a sophisticated and robust algorithm. This will help the intended recipient to recover the original message by means of a cryptographic key. For those who do not have the key, the encrypted message will appear as a stream of meaningless codes (Bender, Gruhl, Morimoto, & Lu, 1996). To overcome the weakness of cryptography, steganographic techniques are suggested to camouflage the presence of the hidden data in such a manner that no one other than the sender and the proposed recipient even recognizes that there is a hidden message (Koppola, 2009). Unlike the other forms of communication, the main purpose of steganography is defeated when the communication between sender and receiver is detected. Therefore, the first requisite of a steganographic structure is its undetectability. In other words, a steganographic system is considered insecure if anyone who isn't in on the secret can differentiate between the cover-objects and stego-objects (Kharrazi, 2006).

A. Yahya, *Steganography Techniques for Digital Images*,
https://doi.org/10.1007/978-3-319-78597-4_1

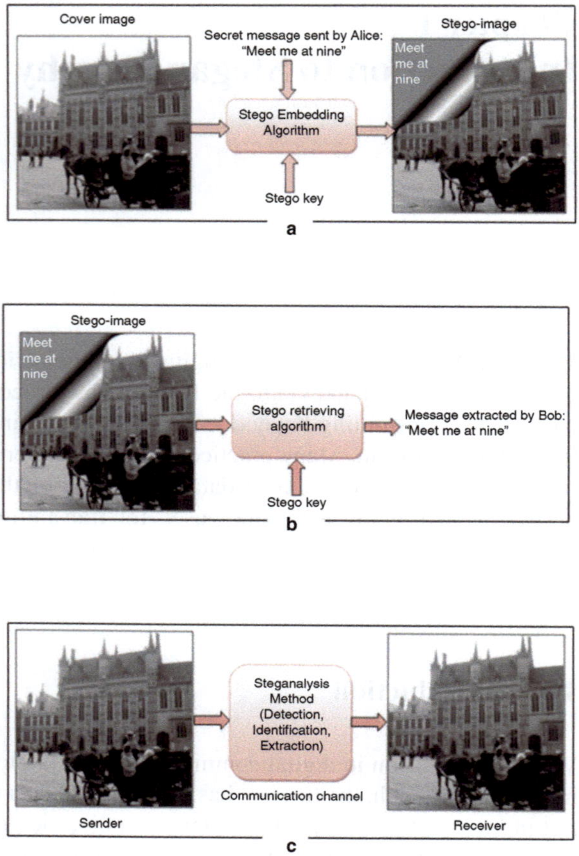

Fig. 1.1 The prisoners' problem (**a**) Steganography embedding system used by Alice (**b**) Steganography retrieval system used by Bob (**c**) Steganalysis system developed by the warden, Wendy

Digital images are commonly used as cover-objects to convey the hidden information. Owing to the high availability of digital images and the high degree of redundancy they showcase, there is a correspondingly high level of interest in the use of images as cover objects in steganography. Much investigation has been described relative to the practices of hiding data in images (Chang, Chen, & Lin, 2004; Chang, Lin, & Hu, 2002; Cheddad, Condell, Curran, & Kevitt, 2009; Chi-Kwong & Cheng, 2001; Hamid et al. 2012b, c, d; Kharrazi, 2006; Koppola, 2009; Ling, 2005; Ran-Zan, Chi-Fang, & Ja-Chen, 2000). Generally, certain terms are used to characterize hidden information. In this book, "cover-image" defines the image that has been designated for hiding the secret data. The term "stego-image" is used for the image that contains the embedded information. In addition, the expression "stego-key" refers to the parameter used to prevent other parties from extracting the secret message from the stego-image. And the processing of an image and the efforts of statistical analysis needed for breaking steganography algorithms are known as "steganalysis" or "attacks."

The classical model for modern steganography was proposed by Simmons in 1984 as "the prisoners' problem" (Simmons, 1984). An example is illustrated in Fig. 1.1

(Katzenbeisser, 2000). In this example, Alice and Bob are in prison and have been thrown into different cells. The two prisoners would like to develop an escape plan; however, all communications between them are monitored by a warden named Wendy, who is known to Alice and Bob as "the adversary." Wendy will not let the prisoners communicate through encryption or any other means that make the communication noticeable. To avoid alerting Wendy of any covert message, an ideal way of communication is used to hide the stego-message within a cover file, such as an image. Figure 1.1 illustrates the prisoners' problem where Alice places a hidden message, "Meet me at nine," and Bob is able to reconstruct the message with a shared stego-key. Note that the difference between the cover-image and the stego-image is visually unobservable. Wendy is unaware that the picture sent by Alice contains the secret escape plan (i.e., the stego-message).

The focus of the current research is on the design of data-hiding techniques used for transmitting secret data where digital images are selected as the cover-media. In the proposed techniques, the emphasis is placed on enhancing the robustness and imperceptibility of the process. Moreover, error-free recovery of the embedded secret data without referring to the original cover-media is required.

1.2 Fundamental Requirements for Steganography

The presentation of a steganographic system can be measured by means of more than a few properties (Cox, Miller, Bloom, Fridrich, & Kalker, 2008):

Imperceptibility (undetectability) of the information measures how difficult it is to control the existence of a hidden message. This parameter is the first and primary requirement; it represents the ability to avoid human-eye detection. Procedures that do not modify the image in such a way as to be traceable by the human eye will possibly still be able to adjust the image in a manner so that it is measurable by the statistical assessments. Accurately secure steganographic methods must be untraceable, either by the human eye or by the statistical threats (Amirtharajan & Rayappan, 2012; Bahi, Couchot, & Guyeux, 2012). To assess the level of imperceptibility, the visual difference between the original cover-image and the stego-image is calculated. By comparing the original cover-image and the final stego-image, the visual difference is determined, and then the imperceptibility level is specified (Ling, 2005).

Robustness refers to how well the steganographic system resists the extraction of hidden data, and the robustness factor measures the capability of the steganographic practice to resist the efforts of eliminating those secret data. These attacks could be the cropping or rotating of the image, noise, image filtering, and data compression (Bahi et al., 2012).

Payload capacity represents the maximum amount of information that can be safely embedded in a work without creating statistically detectable objects. The more data bits to be hidden in the cover-image, the higher the embedding capacity that will be required. In general, imperceptibility is not directly proportional to the embedding capacity, but rather is inversely proportional: when the embedding capacity increases, the imperceptibility level decreases, and vice versa. By using the aforesaid factor, even a minor packet of information could deceive the naked human eye. However, employing statistical analyses or a Human Visual System (HVS) may identify pieces of large data to some extent (Chen, Zhang, Chen, Fu, & Wu, 2012; Lee & Huang, 2012).

Reliability is the most important parameter that characterizes the usefulness of a stego-system based on error-free recovery of the hidden secret information. In other words, a feasible stego-system should allow its users to be able to retrieve hidden information without any loss (i.e., with 100% recovery). In addition to the other three factors, this vital factor should be taken into consideration in designing an applicable and accurate information-hiding system.

It is noteworthy that it is impossible to obtain the highest degree of robustness and the maximum embedding capacity with an acceptable level of imperceptibility at the same time. Therefore, a compromise must be made between robustness, imperceptibility, and embedding capacity. For different applications, the acceptable balance between these three constraints is different, depending on the nature of the requirements of the application (Ling, 2005).

1.3 Steganography Challenges

Ensuring protection and security of long-distance communication is a critical problem. This is particularly important in the case of confidential data storage and transmission on a public network like the Internet. The security of such data communication, which is compulsory and vital for many current applications, has been a foremost concern and an ongoing topic, since the Internet is open by plan and public in nature (Ling, 2005). Countless solutions have been offered for providing the secure transmission of data. Data encryption and information-hiding systems have become prevalent, and they typically accompany each other (Shankar, Sahoo, & Niranjan, 2012). The main problem is that once you encrypt a file, even with a strong encryption algorithm, it looks like a random stream of bytes (Hamid et al. 2012a,b,c,d). In the computer world, random bytes are very rare and therefore very easy to detect in a flow of trillions of structured bytes (Marwaha, 2010). The research work in this book has identified the following problems in the current image-steganography schemes:

- The majority of the existing methods presuppose that flexibility to noise, compression, and other image-processing manipulations are not necessary in the steganographic context, since such manipulations clearly are not custom-made for steganography applications where flexibility, robustness, and security are compulsory (Cheddad, Condell, Curran, & Kevitt, 2010).
- Conventional image-steganography methods alter nearly all the parameters and important components while embedding the secret data. These methods may be more susceptible to attacks and may resultantly increase the chances of losing data, together with degrading image quality (Fallahpour & Sedaaghi, 2007; Koppola, 2009; Widadi, Ainianta, & Chan Choong, 2005).

- Many current steganography schemes hide the information in the predefined locations of an image (Li, He, Huang, & Shi, 2011; Shejul & Kulkarni, 2011). Such an arrangement makes it easy to predict the embedding locations by studying a many stego-images.
- Other steganographic systems intend to embed data in the edges of the image (Chen, Cao, Fu, & Ma, 2011; Jae-Gil, Eun-Joon, Sang-Ho, & Kee-Young, 2008; Singh, Singh, Singh, & Devi, 2007; Hamid et al. 2012a,b,c,d). This method has the advantage of selecting the locations for embedding dynamically. However, it is very image-dependent and has very limited hiding capacity.

Accordingly, a new scheme is proposed, Characteristic Region-Based Image Steganography (CR-BIS), in which the encrypted secret data are embedded in the robust regions of an image. In addition, the extraction process is completely blind, since only the stego-image is required to initiate the extraction process. Besides, an improved robust and secured steganography algorithm (IRSS) is introduced, by taking advantage of the ideas derived and concluded from the CR-BIS algorithm (Hamid et al. 2012a,b,c,d). It will be shown, principally by extensive theoretical studies and comprehensive simulations, that the proposed scheme is significantly better than the existing steganographic schemes.

1.4 Scope of Work

In this book, two different approaches have been provided to solve the problem of securing communication networks. The new approaches are CR-BIS and IRSS.

The first scheme revolves around studying the characteristics of the cover-image to be able to decide the most suitable regions that guarantee robust and secure data transfer. CR-BIS includes multi-layer security and consists of different components (i.e., data encryption algorithm, characteristic regions detection algorithm, and the wavelet embedding algorithm).

By making use of the benefits obtained from the CR-BIS algorithm, the second scheme, IRSS, is developed to improve the performance of an existing scheme (Mali, Patil, & Jalnekar, 2012), which fails to fully recover the data at the receiving end. This makes the scheme unreliable and inapplicable, since the feasible stego-system should ensure a full recovery of the hidden information. Accordingly, an improved steganographic system is proposed in an attempt to successfully embed secret data within the frequency domain by modifying the DCT coefficients. Based on the specific selection criteria, certain blocks have been selected for the concealment of data. To ensure a full recovery of the hidden message, an embedding map has been proposed to indicate the selected embedding blocks. To secure the embedding map, a Speeded-Up Robust Features technique (SURF) has been used to define dynamically the locations in which the embedding map is concealed. In addition, the embedding map has been kept hidden in the frequency domain via modifying the DWT coefficients in a content-based manner.

The characteristics of a steganographic system could be enhanced by adopting some techniques such as data compression. This will in return allow much data to be embedded, or the ECC to enhance the robustness. Accordingly, the last part of this book investigates the effect of ECC and redundancy that can be used to increase the robustness of steganographic systems. For evaluation purposes, each technique is combined with four well-known data embedding methods. Those methods are DCT quantization, DWT quantization, histogram equalization, and DWT content-based, which is the main concern of the newly designed CR-BIS algorithm. As far as the scope of the present work is concerned, the simulation results are expected to be adequate to prove the viability of the new proposed schemes, and their superior performance (as theoretically expected) compared with the other existing schemes. All evaluations and tests are implemented using standard grayscale images (Hamid et al. 2012a,b,c,d).

References

Amirtharajan, R., & Rayappan, J. B. B. (2012). Inverted pattern in inverted time domain for icon steganography. *Information Technology Journal, 11*(5), 587–595.

Bahi, J. M., Couchot, J. F., & Guyeux, C. (2012). Steganography: A class of secure and robust algorithms. *Computer Journal, 55*(6), 653–666.

Bender, W., Gruhl, D., Morimoto, N., & Lu, A. (1996). Techniques for data hiding. *IBM Systems Journal, 35*(3.4), 313–336.

Chang, C.-C., Chen, G.-M., & Lin, M.-H. (2004). Information hiding based on search-order coding for VQ indices. *Pattern Recognition Letters, 25*(11), 1253–1261.

Chang, C.-C., Lin, M.-H., & Hu, Y.-C. (2002). A fast and secure image hiding scheme based on LSB substitution. *International Journal of Pattern Recognition and Artificial Intelligence, 16*(04), 399–416.

Cheddad, A., Condell, J., Curran, K., & Kevitt, P. M. (2009). A secure and improved self-embedding algorithm to combat digital document forgery. *Signal Processing, 89*(12), 2324–2332.

Cheddad, A., Condell, J., Curran, K., & Kevitt, P. M. (2010). Digital image steganography: Survey and analysis of current methods. *Signal Processing, 90*(3), 727–752.

Chen, G., Cao, M., Fu, D., & Ma, Q. (2011, August 16–18). *Research on an Steganographic Algorithm Based on Image Edge.* Paper presented at the Internet Technology and Applications (iTAP), 2011 International Conference on.

Chen, G., Zhang, M., Chen, J., Fu, D., & Wu, Y. (2012). Capacity and security for imperfect batch steganography. *Możliwości i bezpieczeństwo niedoskonałej steganografii pakietowej, 88*(7 B), 324–327.

Chi-Kwong, C., & Cheng, L. M. (2001). Improved hiding data in images by optimal moderately-significant-bit replacement. *Electronics Letters, 37*(16), 1017–1018.

Cox, I. J., Miller, M. L., Bloom, J. A., Fridrich, J., & Kalker, T. (2008). Chapter 12 – Steganography. In *Digital Watermarking and Steganography (Second Edition)* (pp. 425–467). Burlington: Morgan Kaufmann.

Fallahpour, M., & Sedaaghi, M. H. (2007). High capacity lossless data hiding based on histogram modification. *IEICE Electronics Express, 4*(7), 205–210.

Hamid, N., Yahya, A., Ahmad, R. B., & Al-Qershi, O. M. (2012a). An improved robust and secured image Steganographic scheme. *International Journal of Electronics and Communication Engineering & Technology (IJECET), 3*(2), 484–496.

Hamid, N., Yahya, A., Ahmad, R. B., & Al-Qershi, O. M. (2012b). A comparison between using SIFT and SURF for characteristic region based image steganography. *International Journal of Computer Science Issues (IJCSI), 9*(3), 110–116.

Hamid, N., Yahya, A., Ahmad, R. B., & Al-Qershi, O. M. (2012c). Image steganography techniques: An overview. *International Journal of Computer Science and Security (IJCSS), 6*(3), 168–178.

Hamid, N., Yahya, A., Ahmad, R. B, & Al-Qershi, O. M. (2012d, April 10–12). Characteristic region based image steganography using speeded-up robust features technique. Paper presented at the 1st international conference on future communication network (ICFCN'12). IEEE international conference, Iraq, Baghdad.

Jae-Gil, Y., Eun-Joon, Y., Sang-Ho, S., & Kee-Young, Y. (2008, April 7–9). *A New Image Steganography Based on 2k Correction and Edge-Detection.* Paper presented at the Information Technology: New Generations, 2008. ITNG 2008. Fifth International Conference on.

Katzenbeisser, S. C. (2000). Principles of steganography. In S. Katzenbeisser & F. A. P. Petitcolas (Eds.), *Information hiding techniques for steganography and digital watermarking* (pp. 17–41). Boston, London: Artech House.

Kharrazi, M. (2006). *Image steganography and steganalysis.* PhD Dissertation, Polytechnic University. (30250643)

Koppola, R. R. (2009). A high capacity data-hiding scheme in LSB-based image steganography. In *University of Akron.*

Lee, C. F., & Huang, Y. L. (2012). An efficient image interpolation increasing payload in reversible data hiding. *Expert Systems with Applications, 39*(8), 6712–6719.

Ling, L. S. (2005). Study of steganographic techniques for digital images. Master of Philosophy thesis, City University of Hong Kong, City University of Hong Kong.

Li, B., He, J., Huang, J., & Shi, Y. Q. (2011). A survey on image steganography and steganalysis. *Journal of Information Hiding and Multimedia Signal Processing, 2*(2), 142–172.

Marwaha, P. (2010, July 29–31). *Visual cryptographic steganography in images.* Paper presented at the Computing Communication and Networking Technologies (ICCCNT), 2010 International Conference on.

Mali, S. N., Patil, P. M., & Jalnekar, R. M. (2012). Robust and secured image-adaptive data hiding. *Digital Signal Processing: A Review Journal, 22*(2), 314–323.

Ran-Zan, W., Chi-Fang, L., & Ja-Chen, L. (2000). Hiding data in images by optimal moderately-significant-bit replacement. *Electronics Letters, 36*(25), 2069–2070.

Shankar, T. N., Sahoo, G., & Niranjan, S. (2012). Steganographic manipulations with elliptic curve cryptography. *International Journal of Electronic Security and Digital Forensics, 4*(4), 280–297.

Shejul, A. A., & Kulkarni, U. L. (2011). A secure skin tone based steganography using wavelet transform. *International Journal of Computer Theory and Engineering, 3*(1), 16–22.

Simmons, G. (1984). The Prisoners' Problem and the Subliminal Channel. In D. Chaum (Ed.), *Advances in Cryptology* (pp. 51–67): Springer US.

Singh, K. M., Singh, L. S., Singh, A. B., & Devi, K. S. (2007). *Hiding Secret Message in Edges of the Image.* International Conference on Information and Communication Technology, 2007. ICICT '07 Egypt.

Widadi, K. C., Ainianta, P. H., & Chan Choong, W. (2005). *Blind Steganography using Direct Sequence/Frequency Hopping Spread Spectrum Technique.* Paper presented at the Information, Communications and Signal Processing, 2005 Fifth International Conference on.

Chapter 2
Steganography Techniques

Abstract This chapter reviews the main steganographic methods for both lossy and lossless image formats, such as JPEG and BMP. The values are presented in terms of a taxonomy that focuses on three principal steganographic techniques for hiding information in image files. Those practices comprise of those modifying the image in the spatial domain, in the transform domain, and those modifying the image file formatting. Each of these systems tries to satisfy the three most important aspects of steganographic design (imperceptibility or undetectability, capacity, and robustness). It is possible to defeat the transform domain methods by employing some efforts. With most of the steganography applications, the JPEG file format can be used, particularly with images that have to be communicated over an open system environment, like that of the Internet. Thus, for an agent to send secret information using steganographic methods, s/he must select an appropriate steganographic algorithm and cover-image.

2.1 Introduction

Irrespective of distances, in this age of computers and the Internet, people around the world can connect virtually to share data and information. However, threats to confidential and important data have also increased. To overcome these issues, there is a need to design and develop steganography systems. Steganography coupled with encryption is an important means of securing data, particularly when the threat level is high (Cheddad, Condell, Curran, & Kevitt, 2010).

The term *steganography* is often confused with *cryptography* due to some mutual similarities. Nonetheless, they differ from each other on many grounds. Cryptography changes the data shape to maintain secure communication; thus, intruders fail to understand the data. Conversely, steganography methods tend to hide the presence of the data, thus making it impossible for spies to read and steal the private embedded information. In some circumstances, transmitting encrypted data may be more vulnerable, whereas hidden message are not. Thus, cryptography is not the only or best solution to secure information from external threats (Abdul-mahdi et al., 2013).

A. Yahya, *Steganography Techniques for Digital Images*,
https://doi.org/10.1007/978-3-319-78597-4_2

Fig. 2.1 The Al Jazeera TV channel's visible watermark

The combination of these two sciences can overcome a threat more efficiently. In a scenario where steganography fails, the information or data still cannot be retrieved when the message is also encrypted through cryptography (El-Emam, 2007).

To keep intellectual property safe, watermarking and fingerprinting methods coupled with steganography are commonly employed (Morkel, Eloff, & Olivier, 2005). In a digital watermark, the signal is permanently embedded into digital data (e.g., audio, video, in-text images), but is discoverable to validate the statistics. The watermark is embedded in the host statistics so that it cannot be extracted without harming the host medium. Nonetheless, this technique keeps the information accessible (Lu, 2005). The invisible data in a watermarked item is a monogram, documenting the real possessor of the data to certify copyright protection. Figure 2.1 displays the watermarked logos of the Al Jazeera TV channel for their transmission. In the scenario of fingerprinting, some definite and various signs are embedded into replicas of a work for different clients. In such a situation, it is simple for the possessor of intellectual property to recognize such clients who provide themselves the authority to disrupt their licensing contract when they transfer the property to others unlawfully (Anderson & Petitcolas, 2006). Figure 2.2 presents the key divisions of the security scheme.

Practically, all digital file formats are recognized for their utilization for steganography, with an extraordinary mark of redundancy. The redundant portions denote those chunks with the ability of being altered without any option to sense the modification, such as image and audio files (Morkel et al., 2005). Actually, digital images are the carrier file setups commonly chosen due to their reputation on the Internet. The current research is confined to steganography using image file formats. Video, audio, and linguistics are other categories of steganography, but they are beyond the scope of this book.

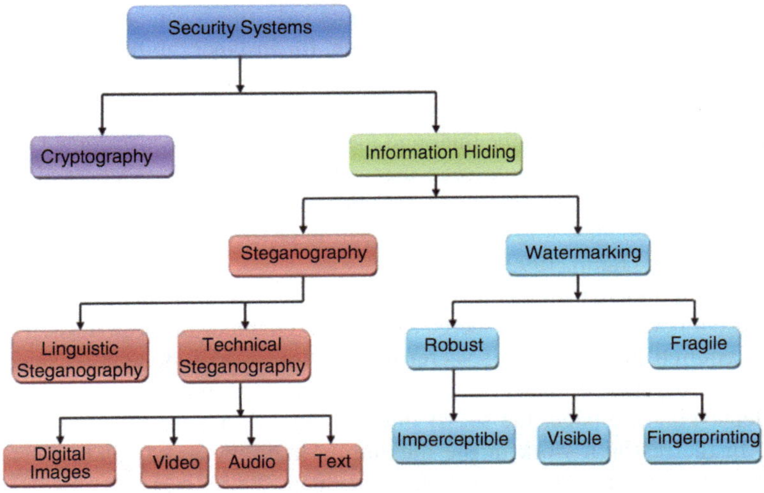

Fig. 2.2 Security system branches. (Cheddad, 2009)

This chapter reviews the literature and background study of steganography; it provides a classification of the existing steganographic methods for image file formats. These methods are investigated and deliberated not only in terms of their capability to hide data in image files, but also their toughness against diverse image processing threats and the quantity of data which can be hidden. Lastly, this chapter gives a comparison analysis that attempts to conquer and disrupt diverse steganography algorithms; steganalysis.

2.2 Background and History of Steganography

In the past, people used to hide data and information by employing various means. The history of steganography is equally profound as the past itself. The term "steganography" came fundamentally from the Greek term that stands for "covered writing". Later, the term was used by scientists for thousands of years in numerous ways (Hamid et al., 2012e,f; Provos & Honeyman, 2003). In the fifth century BC, the Greek dictator Histiaeus was imprisoned by King Darius in Susa, who wanted to transmit complex messages to his son-in-law, Aristagoras, in Miletus. To send the secret message, Histiaeus used the scalp of his slave's head to tattoo the secret information. He was waiting for the slave's hair to cover the tattoo and, afterwards, he was sent to Miletus to deliver the secret message (Johnson, 1995). In the earliest Greece, peeling the wax off a wax-covered tablet was one of the techniques used to read and write secret messages. The recipient of the message would merely dispose of the wax from the tablet to read the hidden message.

Unseen ink was one of the prevalent methods of steganography. Early on, Romans used unseen ink and other materials, for instance, milk, urine, and fruit juice, to write secret messages between the lines. Although the usage of unseen or invisible ink appears not to be hurtful, a letter might reproduce very diverse information printed between the lines and the same technique was adopted in World War II to transmit a secret message using invisible ink (Cheddad et al., 2010). Furthermore, the Germans employed the microdot method throughout World War II. Data, mainly images, was made very small in order to make it complicated and difficult to detect (Jamil, 1999).

In 1550, an Italian mathematician, Jerome Cardan, used a paper mask with holes to propose a new technique of secret writing. The handler of these mask papers is merely required to compose his secret information in such holes after masking a blank piece of paper. In the subsequent phase, the blank portions are filled in and, after demasking the page, the secret information will appear as safe text (Jamil, 1999). In the research work in the literature (Hamid et al., 2013a,b; Johnson & Jajodia, 1998; Judge, 2001; Provos & Honeyman, 2003), a broad background of steganography can be discovered.

Steganography is considered the most sophisticated and robust dispatch channels technique widely used in computer digital data. The subsequent section demonstrates the classification of steganographic methods for image files, as well as covering an impression of the most significant steganographic systems in digital images.

2.3 Steganography Practical Applications

There are number of valuable applications of steganography, such as enrichment of the strength of image search engines and smart identity cards coupled with information of individuals embedded in their photos and the copyright control of materials. Other important applications comprise video–audio synchronization, the secure movement of companies' secret information, television transmission, and TCP/IP protocol packets, where distinctive data are embedded in an image for the persistence of investigating the network traffic of specific handlers (Johnson & Jajodia, 1998). On this matter, Petitcolas (2000) presented some advanced applications, such as medical imaging systems, one of the most important applications from his research work. The author recommended a separation between patients' image data or DNA structures and their descriptions for privacy motives, such as doctors' and patients' demography. Henceforth, embedding the patients' data in the image might be a valuable protection method and benefits in resolving such difficulties.

Thus, steganography supports the delivery of a decisive assurance of verification that no further safety apparatuses can offer. In this way, an embedding system that inserts electronic patient records through a bi-polar multiple-base message embedding method is presented by Shaou-Gang, Chin-Ming, Yuh-Show, and Hui-Mei, (2000). Moreover, the impression of masking patients' records in digital images are presented and centered on the concept that steganography can be measured as a usual printing procedure (Anand & Niranjan, 1998; Yue, Chang-Tsun, & Chia-Hung, 2007). Fujitsu, a famous Japanese company, established a system to encrypt information into a printed image that cannot be seen by the human naked eye. Conversely, the image can be easily decrypted by a camera built into a mobile phone. This practice benefits from the widespread use of mobile phones in handling the encrypted information (Cheddad et al., 2010).

Digital expertise has depended on its self-reliance in the reliability of graphical imagery (Farid, 2009), a problem that has inspired investigators to conduct studies on digital file forensics. Cheddad and his co-researchers suggested a steganographic structure which protects scanned files from counterfeiting by employing methods which are self-embedded. The scheme permits legal or forensics professionals to have right of access to the real file, even if it is forged (Cheddad, Condell, Curran, & McKevitt, 2009a,b).

2.4 Applicable Image Steganography Methods

In this division, an outline of steganographic methods focusing on precise image formats is presented. An image is defined as a prearrangement of figures which represents diverse light strengths in various portions of the image. The numeric explanation takes the type of a frame where each point is specified by the term "pixel". The number of bits in a color scheme is recognized as the bit gravity and this mostly denotes the number of bits allocated to each pixel (Morkel et al., 2005). The graphics interchange format (GIF), joint photographic experts group (JPEG), and, to a lesser degree, the portable network graphics (PNG) are considered the most common image formats existing on the Internet. In many cases, steganographic methods try to manipulate the basic configuration of these formats. Nonetheless, in some research work, the bitmap format (BMP) has been used due to its non-complex data structure and simplicity (Cheddad et al., 2010). Image steganographic methods are categorized based on different techniques in accordance with the category of concealment employed with classified communications. Alternatively, it is performed by arranging such methods subject to the category of cover adjustment previously utilized during the procedure of embedding. The second method is implemented for the classification offered in chapter three, while, in some circumstances, a precise arrangement is impossible. Largely, the graphical procedure of embedding can be clarified as presented in Fig. 2.3.

Let C represent the cover carrier and $C^{\sim}$ denote the stego-image. Let K stand for a seed employed to encode the data or to produce a pseudo-random noise, which can be set to $\{\varnothing\}$ for simplicity, and let M represent the data to be transmitted. Then, E_M and E_X will signify the embedded and extracted data, respectively. Consequently,

$$E_m : C \oplus M \oplus K \rightarrow C^{\sim} \tag{2.1}$$

$$\therefore E_X\left(E_m\left(c,m,k\right) \approx m, \forall c \in C, m \in M, k \in K \tag{2.2}$$

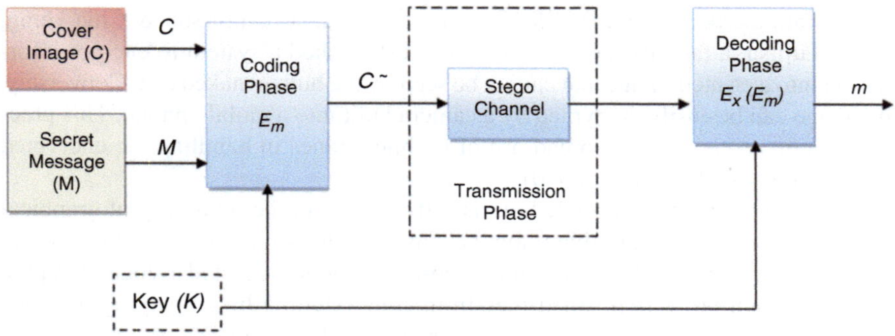

Fig. 2.3 Steganographic system model

To differentiate amongst the diverse steganographic methods in an extensive manner, it is important to note the role of both techniques: the one that alters the image and the other that changes the image file format. Nevertheless, the alterations to the file format are non-robust (Kruus, Scace, Heyman, & Mundy, 2003). The vital matter to declare here is the key part of compression, which typically shows when there is confusion and competition to choose the best steganographic algorithm. Yet, lossy compression approaches result in tiny image document sizes, since the chance of partial loss of embedded data increases. The reason behind these techniques is the exclusion of additional data during processing. Lossless compression usually does not compress the image document greatly (Prasad, Janeyulu, Krishna, & Nagaraju, 2009). As a result, scientists have come up with new and modified steganographic algorithms suitable for data compression. The following are steganographic systems that alter image documents for invisible data:

- Spatial domain;
- Transform domain;
- Spread spectrum;
- Statistical methods; and
- Distortion techniques.
- Steganographic schemes that adjust the image document format include document and palette embedding.

Furthermore, there are systems that alter the elements in the graphic image. The latter comprise image generation and image element alteration schemes. Lastly, the distinctive category of the spatial and transform domain scheme, known as adaptive steganography, characterizes the core issue of the existing study. The following sections exemplify the classification of steganographic schemes for image documents, as well as giving an outline of the steganographic systems in digital images of utmost significance.

2.4.1 Spatial Domain Steganographic Systems

Spatial domain steganographic systems is a collection of comparatively easy methods likewise recognized as replacement methods. These schemes support the generation of a hidden channel in the portions of the cover image, where variations are likely slightly restricted once related to the HVS. One way to achieve the desired result is to hide and embed the data in the least significant bit (LSB) of the image data (Kruus et al., 2003). This technique of hiding important information is solely dependent on the statistic that the LSBs in an image can be assumed to be random noise, and, subsequently, non-responsive irrespective of any alterations to the image (Chandramouli, Kharrazi, & Memon, 2004).

There are two systems, sequential and scattered, which are used to hide the information bits in the image using LSB algorithms. The first is demonstrated by the LSBs of the image in the sequential embedding system, which are substituted by the information bits, while the scattered embedding system includes the information bits to be randomly scattered throughout the image by means of a random sequence to regulate the embedding sequence (Juneja & Sandhu, 2009).

The renowned steganographic methods established on LSB embedding are diverse as far as the way in which they embed data is concerned. Resultantly, the LSB of pixels are altered by these techniques randomly; some alter pixels in selected areas instead of the whole image and others escalate or reduce the pixel value of the LSB, instead of altering its value (Cheddad, 2009). A number of steganographic techniques employing the LSB have been established in the steganographic system; for instance, Steghide, S-Tools, Steganos, etc. are offered on the Internet (Johnson, 2009). These methods can accomplish a high capacity; however, they do not offer robustness besides an easy modification on stego-images and are easily identified. Numerous variations on the simple LSB methods have been defined by Johnson and Katzenbeisser (2000). The authors also defined a replacement method for hiding secret data or information in the LSBs of the palette of the GIF or BMP image formats by employing steganography. However, the authors did not include test images that can permit the readers to imagine the ideas in practice. They complete their investigation without future references or improvements.

A team of scientists upheld that, regardless of the statistic that switching the LSB of one pixel in a JPEG image acquires a small modification (Fridrich, Goljan, & Hogea, 2003), this modification can still be discovered. These trials on discrete cosine transform (DCT) coefficients verify encouraging outcomes and attract scientists' attention concerning this category of images. Conversely, the tangible stand-in at the level of DCT marks steganography to not be as suitable against statistical threats and needs to be more resilient (Hamid et al., 2013a,b).

In the work by Wu and Tsai (2003), the authors proposed an innovative steganographic method where the number of secret bits for an embedded message depends on the difference value between two neighboring pixels. The authors divided a cover image into non-coinciding blocks of two successive pixels. In their scheme, the authors calculated the difference value from the two pixels values in each block.

In the next phase, all likely difference values are gathered into a number of series. The choice of series interval is established based on the features of the human vision toward gray value disparities from evenness to divergence. The difference value at that moment is substituted by a fresh value to insert one value of the important secret data. This process delivers a simple method to produce indiscernible outcome as compared to the simple LSB substitution scheme. It is noted here that the pixel-value differencing (PVD) steganography can hide and embed a huge quantity of secret bits into images with great indiscernibility by reason of its aptitude to utilize the features of human vision. In contrast, extraordinary phases in the histogram of pixel modifications disclose the presence of embedded secret data.

A side match method is planned to embed and hide secret information, with the aim of delivering greater hiding volume and to decrease the stego-image distortion. The difference between the pixel and its upper and left side pixels identify the number of bits to be embedded in a pixel. The proposed scheme performs the correlation between the adjacent pixels to evaluate the notch of evenness or the disparity of pixels. Specifically, if the pixel is positioned in a brink zone, it may withstand greater deviations as compared to those in even zones (Chang & Tseng, 2004). The difference between the pixel and its upper and left side pixels identify the number of bits to be embedded in a pixel. This approach will bound the hiding capacity, subject to the image quality being employed as a cover image. Deeper research is required for the cover image to obtain the best outcomes. It has been proved from their study that these techniques are not appropriate for all images.

In 2005, a hybrid steganographic system was planned by merging both PVD and LSB substitution to hide and embed important secret information into still images (Wu, Wu, Tsai, & Hwang, 2005). The authors used the method where the tendency to embed secret information into edged areas is greater as compared to smooth areas in the cover image and has improved image quality as a result of using the PVD technique only. The PVD technique enhances the capability of the system for the reason that the important secret information is better hidden and embedded in the even zones by employing an LSB scheme. The experimental outcomes disclose that the planned system results in stego-images with a suitable value and offers huge embedded important secret information capability (Wu et al., 2005). However, there are some deficiencies in their methodology which were not addressed properly. For instance, the combination of PVD and LSB schemes was not compared to the LSB method in the embedding process. Also, in cooperation with LSB, the combination of PVD and LSB schemes is more vulnerable, as mentioned by Fridrich, Goljan, and Rui (2001).

In 2008, Wang, Wu, Tsai, and Hwang (2008) proposed a novel steganographic system by making the best use of the remainder of two consecutive pixels to secure the secret information with extra efficiency to be capable of developing the optimum remainder of the two pixels at the minimum conceivable deformation. By way of employing this structure, the embedding influence in the stego-image seems to be able to be expressively compact in contrast with the system offered in the work of Wu and Tsai (2003). Experimental outcomes from the work of Wang, Wu, Tsai, and Hwang (2008) show that the projected structure outperforms the system developed by Wu and Tsai (2003) in terms of stego-image excellence. Nevertheless, the structures of edge embedding are not taken into account by a number of scientists (Wang et al., 2008; Wu et al., 2005; Wu & Tsai, 2003). In the meantime, some new techniques have been developed to address the aforesaid issue; however, these new schemes result in broadcast errors, coupled with lower embedding capability (Chang & Tseng, 2004; Park, Kang, Shin, & Kwon, 2005).

In 2011, Liao, Wen, and Zhang (2011) proposed an innovative steganographic system to deliver improved stego-image superiority, coupled with greater embedding capability, to enhance the multi-pixel differencing system, grounded on a modified LSB substitution. A system having a four-pixel block with three difference values is employed, the same as that designed by Ki-Hyun, Kyeoung-Ju, and Kee-Young (2008) and Yang and Wang (2006).With the purpose of differentiating between the edged and even zones and, correspondingly, to evaluate the quantity of secret bits that are able to be embedded into the block, an average assessment of three difference values has been employed. Afterwards, by implementing a modified LSB substitution system, secret bits are embedded easily into individual pixels in the block. Their planned methodology emphasized the structures of edges specifically, so that the pixels in edge zones are able to accept additional deviations, short of creating distinguishable distortion (Liao et al., 2011). Resultantly, the projected system attained adequate capabilities and made it undetectable. Conversely, still at hand is an exchange between the embedding capability/excellence and threat confrontation. Moreover, the projected scheme trades threat resistance confrontation for attaining complex embedding capability/excellence.

It has been established that the employment of LSB systems make it impossible for the naked human eye to document the subsequent fluctuations to the cover image, due to them being very small. Besides, such procedures are guileless and prevalent. In this procedure, each and every pixel in the image is being used, which is the biggest shortcoming of the system. Resultantly, the chances of losing hidden data are high if lossy compression is being used. It is an open secret that the development of LSB embedding systems is a great achievement; its property of negligible confrontation to the threats compelled scientists to take advantage of this property and test it in other applications, for instance, the frequency domain, etc. (Cheddad et al., 2010; Geetha, Kabilan, Chockalingam, & Kamaraj, 2011; Noda, Niimi, & Kawaguchi, 2006). The next section clarifies data embedding in the image transform domain.

2.4.2 Image Frequency Domain Steganographic Scheme

A large number of algorithms for the transform domain embedding system have been recommended. The procedure of embedding data in the frequency domain of a signal is amply resilient as compared to embedding systems that function in the time domain. Currently, almost all robust steganographic classifications work within the transform domain (Hamid et al., 2012b, 2013b; Johnson & Katzenbeisser, 2000).

16	11	10	16	24	40	51	61
12	12	14	19	26	58	60	55
14	13	16	24	40	57	69	56
14	17	22	29	51	87	80	62
18	22	37	56	68	109	103	77
24	35	55	64	81	104	113	92
49	64	78	87	103	121	120	101
72	92	95	98	112	100	103	99

Fig. 2.4 Standard quantization table

These systems take the lead from LSB methods for the reason that they hide and embed data in zones of the image with a reduced amount of visibility to compression, cropping, and image processing. Approximately, transform domain procedures do not appear reliant on the image format and they may possibly outclass lossless and lossy format conversions (Kruus et al., 2003). Stego systems in the transform domain hide information in the coefficients of the signified domain, for instance, discrete Fourier transform (DFT), DCT, and discrete wavelet transform (DWT) (Johnson & Jajodia, 1998; Johnson & Katzenbeisser, 2000). In the two-dimensional DCT phase, each 8×8 non-overlapping block is converted into the DCT domain by means of the two-dimensional DCT (Kekre, Mishra, Shah, Shah, & Thakkar, 2012), as given by:

$$F(u,v) = \frac{c(u)c(v)}{4} \sum_{i=0}^{7} \sum_{j=0}^{7} \cos\left(\frac{(2i+1)u\pi}{16}\right) \cos(\frac{(2j+1)v\pi}{16} f(i,j) \qquad (2.3)$$

where $c(e) = \begin{cases} \frac{1}{\sqrt{2}}, if \ e = 0 \\ 0, if \ e \neq 0 \end{cases}$

$F(u,v)$ and $f(i,j)$ represent a DCT coefficient at the (u,v) coordinate and a pixel value at the (i,j) coordinate, respectively. $F(0,0)$ denotes the DC component, which links to an average concentration rate of individual blocks in the spatial domain. $F(u,v)$ denotes the AC component, in which $u \neq 0$ and $v \neq 0$. For data reduction during the quantization phase, DCT coefficients are quantized by using the standard quantization table, as shown in Fig. 2.4. It has been noticed in Fig. 2.1 that it will be difficult for a system to make a large modification, since the upper left values in the quantization table are excessively small. On the contrary, the lower right values in Fig. 2.1 are sufficient to be transformed. This importance of being able to advance

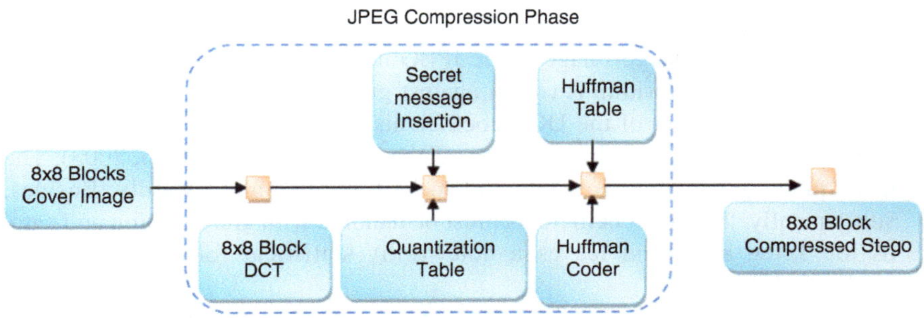

Fig. 2.5 DCT-based steganography system

to camouflage the HVS is the reason that the latter is extra-subtle to the values in low-frequency components as compared to the ones in the higher frequencies. As a result, alteration in high-frequency components is visually suitable and undetectable (Lin, 2012; Lin & Shiu, 2010). Even though embedding at the DCT level is an effective and controlling means, if the coefficients are not cautiously designated, more or less objects, as a result of data embedding, will be evident.

When DCT is employed for JPEG images, the pixels are able to transform with the calculated measurements as shown in Eq. 2.3. In the subsequent phase, the quantization stage of the compression is calculated. Moreover, it is reflected as a natural step where the naked human eye is enacted. Essentially, it is the God gifted property that the human eye is recognized for categorizing minor changes in illumination over a comparatively large zone. A similar concept is not able to be implemented after allowing for the dissimilarity between diverse strong suits in high-frequency intensity (Morkel et al., 2005). As a result, the strong suit of frequencies at higher nodes is able to diminish short of any modification in the image form. The JPEG image format is completed by distributing all the values in a block through a quantization coefficient; thus, the outcomes are estimated to the integer values. Lastly, encryption of the coefficients by employing Huffman coding merely minimizes the size. Earlier, it was understood that steganography possibly will not be employed with JPEG images due to the lossy compression that results in portions of the image information being transformed. Digital cameras, scanners, and other photography tools commonly use the JPEG image format. The reason for this is that JPEG images offer enhanced camouflaging as compared to other formats. In steganographic classifications, information is embedded into the non-zero DCT coefficients of JPEG images. Figure 2.5 displays a steganography established on DCT.

The main JPEGsteganographic approaches are as follows:

- JSteg. This is a typical JPEG steganographic model, which embeds secret data in the transform domain employing the DCT scheme (Johnson & Katzenbeisser, 2000). In this method, LSBs of non-zero quantized DCT coefficients swap with secret data bits to embed the secret data in a cover image. The function of the software is to embed the important secret information into the lowest-order bits of all non-zero frequency coefficients, which result after the DCT has been performed. The outcomes are encrypted using the Huffman coding method to give the end result stego-image in the JPEG image format (Kruus et al., 2003). Even though the algorithm is robust against graphical threats, investigating the statistical distribution of the DCT coefficients displays the presence of hidden embedded information (Cheddad et al., 2010; Provos & Honeyman, 2003). The Chi-square (χ^2) method is used to identify JSteg straightforwardly. Besides, subsequently, the DCT coefficients must be handled with great caution and care, as the JSteg algorithm leaves a noteworthy statistical footprint (Cheddad et al., 2010).

- F5. Westfeld (2001) presented the F5 steganographic algorithm, which embeds information into the non-zero AC DCT coefficients by decreasing their entire values by 1 in lieu of substituting the LSBs of the quantized DCT coefficients with the data bits. The author emphasized that the χ^2 threat is not responsible for this and can never sense this kind of embedding (Westfeld & Pfitzmann, 2000). Later, another steganalysis system based on the natural distribution of DCT coefficients was recommended by Fridrich & Goljan (2002).Their recommended system does identify F5 objects by distracting the F5's existence.
- OutGuess. A UNIX source code system, OutGuess was designed by Provos (2001) with two available types/versions. The primary one is OutGuess-0.13b that shows statistical analysis, while the subsequent one is OutGuess-0.2, which enables protection aptitude towards statistical properties. Henceforth, OutGuess leads to OutGuess-0.2. The embedding development of the OutGuess system is characterized into two phases. The primary phase includes secret data bits embedding alongside a haphazard pace into the LSBs of the quantized DCT coefficients, while avoiding 0s and 1s. The following phase demands nearly the same adjustments, ending with the coefficients previously discovered for the period of embedding. This supports the stego-image comprehensive DCT histogram to match the cover image. OutGuess cannot be exposed to χ^2 threat. The developers of OutGuess offered a counterattack in contrast to the proposed algorithm, which contains a comprehensive form of the χ^2 experiment to choose pseudo-randomly embedded data in JPEG images (Provos & Honeyman, 2003).
- MB. The model-based steganography (MB) technique employs a statistical model of the cover media by means of an all-purpose framework for directing both steganography and steganalysis (Sallee, 2004). The MB scheme for JPEG images has the capability of handling an extraordinary amount data, while continuing to be secure, in contrast to numerous first-order statistical threats (Lin & Shiu, 2010).
- YASS. This is Yet Another Steganographic Scheme from the family of JPEG steganography. However, this system does not cover information in JPEG DCT coefficients openly (Solanki, Sarkar, & Manjunath, 2007). As an alternative, an input image in the spatial domain is distributed into blocks with a secure large size, named big blocks (or B-blocks). Inside each B-block, an 8×8 sub-block, recognized as an embedding host block (or H-block), will be randomly designated. Afterwards, secret information is encrypted and embedded in the DCT coefficients of the H-blocks by employing error correction codes. To conclude,

the whole image is compressed and circulated in the JPEG image format after inversing DCT coefficients on the H-blocks (B. Li et al., 2011).

To the extent that DWT is related, steganography systems in the transform domain aim to embed and hide information by adjusting the DWT coefficients. Wavelets are employed in the image steganographic model for the reason that the wavelet transform evidently screens the high- and low-frequency data on a pixel-by–pixel basis. The DWT technique is preferred over the DCT system, due to the firmness that the DWT technique delivers to the image at numerous stages (Reddy & Raja, 2009).

A team of scientists developed a steganography system that supplements secret data into a base layer transmission of a zero-tree established wavelet coder (Syed, 1999; Syed & Rao, 1999). The authors planned to embed the data in the inappropriate offspring of the detail aspect of sub-carriers in non-smooth sections of the image. The homogeneous connected-region interested ordered transmission (HC-RIOT) coder is employed to decide which section of the image is to be used to embed the data. Correspondingly, it will choose the wavelet coefficients in the detail aspect of sub-carriers of these sections which will possibly be improved. The core point of this method is its capability to transmit steganographic data in lossy situations with sturdiness against threats (Areepongsa, Kaewkamnerd, Syed, & Rao, 2000; Areepongsa, Syed, Kaewkamnerd, & Rao, 2000).

Paulson (2006) stated that a team of researchers at Iowa State University established a cutting-edge application named artificial neural network technology for steganalysis (ANNTS) by means of sensing all existing steganography systems in the image transform domain, together with DCT, DWT, and DFT (Paulson, 2006). ANNTS employed neural networks to statistically scan digital files and, at that point, test them for variations in pixel values so as to spot the accessibility of steganographically embedded hidden data. For instance, this method detected that, in an image without steganography, the amount of pixels with even standards, centered on their location on a gray or color scale, and the amount with odd standards will become dissimilar. In an image with steganographically embedded information, the statistics are nearly alike. This offers a simple statistical check for steganography. It has been concluded that, from the time when the DFT contains a rounding off error, as a result of the reproductions achieved in the DFT with secure point arithmetic, this marks DFT as unfitting for steganography uses.

An information embedding and hiding system in the DWT domain was effectively executed by Abdelwahab and Hassaan (2008). In their system, secret and cover images are together disintegrated by employing one-level DWT, while individuals are distributed into splits of 4 × 4 blocks. At that point, an evaluation is completed concerning the blocks of the secret image and the cover blocks to define the finest counterpart. Far ahead, error blocks are formed and embedded into the coefficients of the finest corresponding blocks in the horizontal sub-carrier of the cover image. Two sources need to be interconnected; the primary one is to clamp the directories to the corresponding blocks in the cover estimated sub-carrier, and the subsequent one stands for the corresponding blocks in the horizontal sub-carrier of the cover. The offered system displays extraordinary strength in contrast to countless image-handling processes, for instance lossy compression, blurring, cropping, median filter, sharpening, coupling with noise. The key shortcoming of this structure is that the removed payload is not completely alike the embedded form. This is by reason of the circumstance that the individual embedded and extracted bits have their place in the secret image estimation once the information is placed entirely in further sub-images to 0s throughout the restoration procedure.

A year later, a steganography system, HCSSD, with extraordinary capability and great security was offered employing DWT (Reddy & Raja, 2009). In the HCSSD scheme, the cover and payload are standardized and the wavelet coefficient is attained by relating a DWT. The estimation band coefficient of the payload and wavelet coefficient of the cover image is merged, established on the strong point of the two parameters, such as alpha and beta. The cover and payload are pre-processed to decrease the pixel series to confirm that the payload is improved precisely at the end point. The capability of the offered algorithm is amplified, as merely the estimation band of the payload is measured. The entropy, mean squared error (MSE), is enhanced with an adequate graphical feature for the subsequent stego-image. On the other hand, there was an oversight of the authors in terms of an argument around the strength of the recommended structure compared to threats.

In 2011, Shejul and Kulkarni (2011) executed an encouraging steganography scheme by means of embedding information inside the covering section of an image. It was supposed that such an embedding offers an exceptionally secure position for information hiding. In their study, biometric steganography was offered by means of the covering section of images in the DWT domain for secret information embedding. Secret information is embedded in some of the high-frequency sub-carriers of DWT by drawing covering pixels in those sub-carriers. The security of the system is improved with or without cropping, and histogram-based threats are effectively prohibited by hiding information in definite sections of an image. The recommended methodology offers an acceptable image value. Alternatively, embedding the secret data in precise sections in the image bounds the payload capability. Furthermore, selecting the covering section of an image for the persistence of hiding will possibly supplement additional restraint for the assessment of images that have been employed as cover images.

For a better understanding of the concept, several authors explain steganography in the DWT domain with more facts and illustrations (Chen, 2007; Potdar, Han, & Chang, 2005; Shen, Zhang, Feng, Cao, & Huang, 2007).

2.4.3 Spread Spectrum Image Steganography (SSIS) Technique

Spread spectrum (SS) modulation tools were established in the 1950s in an effort to deliver a means of low-probability-of-intercept and anti-jamming communication. The SS method is well defined as follows (Hamid et al., 2011a,b, 2012a,b,c; Pickholtz, Schilling, & Milstein, 1982):

> SS is a means of communication wherein the signal dwell in a bandwidth in addition of the least required to send the data; the band spread is accomplished through a code which is independent of the data, and a harmonized reception with the code at the receiver is used for dispreading and successive data retrieval.

Even though the power of the transmitted signal will possibly be large, the SNR will be low in each frequency channel. In the SS modulation scheme, adequate data must be available in the other channels in order to recuperate the signal, even though portions of the signal are detached in numerous frequency channels. Hence, the spread spectrum modulation scheme makes it tough for intruders to identify or interrupt the signal. The case is the same for the steganography scheme; it attempts to spread secret data over a cover image so as to make it difficult to observe (Johnson & Katzenbeisser, 2000). Subsequently, the power of the embedded hidden signal is considerably lower as compared to the power of the cover image; the embedded data becomes undetectable not only to the naked human eye, but also to computer analysis as well without retrieving the original image (Marvel, Boncelet Jr., & Retter, 1999). The spread spectrumimage steganography (SSIS) technique is an information-hiding communication steganographic scheme that employs digital imagery as a cover signal. It delivers the aptitude to hide an important amount of data bits inside digital images and evades being sensed by an observer. The data are recuperated with a low error probability with the help of error control coding (Marvel, Retter, & Boncelet Jr., 1998b). The general additive embedding system can be designated as follows (Marvel, Retter, & Boncelet Jr., 1998a):

$$Y_i = X_i + \gamma W_i \quad \text{for } i = 1, 2, \ldots\ldots, N \tag{2.4}$$

where X_i denotes a sequence of the original data from the cover, W_i represents a pseudo-random sequence produced from a pseudo-random number generator (PRNG) that adjusted in order by a secret stego key, γ represents an embedding strength parameter (gain factor), whereas Y_i is a sequence of the probably changed information.

The main key of SSIS is a spread spectrum encoder (SSE). These policies work together by means of modulating a narrow-band signal above a carrier. The carrier's frequency is repeatedly moved by means of a pseudo-random noise generator coupled with a secret key. In this way, the spectral energy of the signal is spread over a wide band; resultantly, its density, which is typically below the noise level, decreases. In this system, the receiver needs to apply the identical key and noise generator and adjust the exact frequencies to demodulate the original signal in order to remove the embedded data. It is not an easy task for the viewer to identify the embedded hidden communication due to the high noise level (Marvel et al., 1999). A basic SSIS encoder and decoder is shown in Figs. 2.6 and 2.7, respectively. For the duration of the encoding phase, the data is transformed to pseudo-noise and, afterwards, is added to the cover image to produce the steganographic image (clipping and digitizing

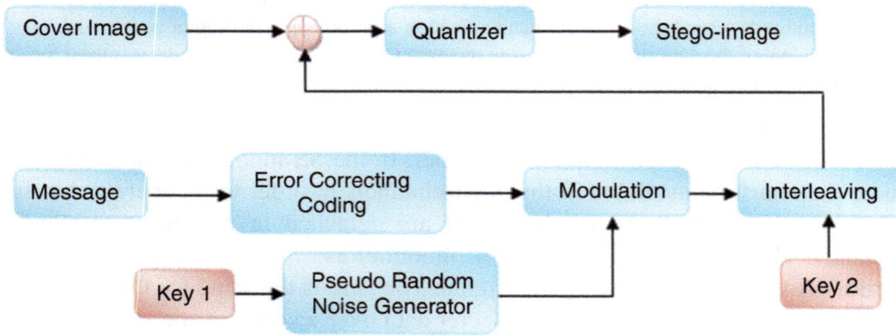

Fig. 2.6 Basic SSIS encoder

Fig. 2.7 Basic SSIS decoder

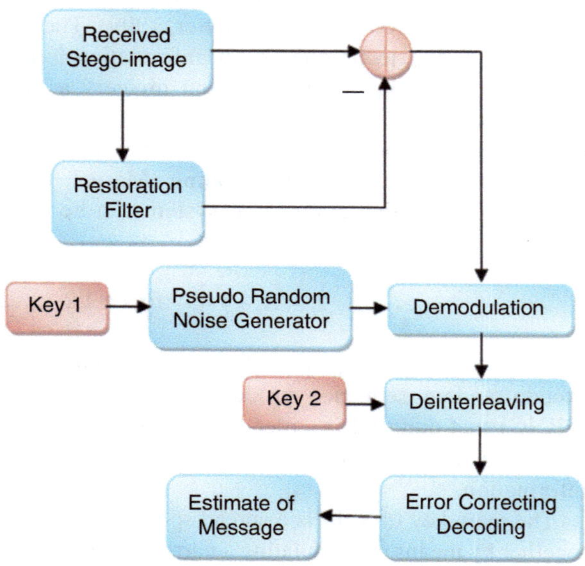

are recommended to generate the digital steganographic image). In order to recover the secret data, initially, the image is filtered to determine the pseudo-noise and, subsequently, the data are removed from the pseudo-noise (Marvel et al., 1999).

In 1996, Smith and Comiskey (1996) recommended some new communication ideas to describe data hiding systems. The authors proposed three systems; explicitly, the direct sequence spread spectrum (DSSS), frequency hopping spread spectrum (FHSS), and chirp. In DSSS, the secret signal is spread by a pseudo-noise code with a constant rate, named the chip rate. Conversely, in FHSS, the frequency of the carrier signal is changed by jumping from one channel to another. The chirp is a hybrid classification, a combination of FHSS and DSSS, typically for the purpose of hiding data. It has been established from the tests that the proposed method coupled with

FHSS is far better than the DSSS as far as its robustness, throughput, and mobility are concerned. However, the DSSS is far better when it comes to computational complexity, which makes it less vulnerable to external threat. Overall, the robustness of such a system remains to be ascertained. A noise filter, comparable to that employed in the SSIS decoder as shown in Fig. 2.7, will eliminate the data exclusively.

To enhance the payload capability, SS practices are able to couple with the transform embedding method by means of transformation methods. A method based on DFT is described by Alturki and Mersereau (2001). This method embeds the secret data in the DFT domain after permuting the image pixels in the spatial domain. The procedure of permuting the image pixels enhances the chances of placement into the cover image. It further enhances expressively the transform coefficients employed for the transmission of secret data (Alturki & Mersereau, 2001).

In 2005, Widadi, Ainianta, and Chan Choong (2005) suggested a blind image steganography, established on a hybrid DSSS/FHSS method. The suggested blind spread spectrumimage steganography (BSSIS) is able to pull through the hidden data from the stego-image without mentioning the original cover image. The BSSIS system is achieved through the supplement of quantization practice and channel estimation. The quantization procedure happens on the encoder side, whereas the channel estimation is performed at the decoder end, like the situation initiated in the core system recommended by Marvel et al. (1999).

The issue of hiding data in a digital host image by using a spread spectrum embedding method in an arbitrary transform domain was deliberated by Gkizeli, Pados, and Medley (2007). This method involves employing the minimum-eigenvalue eigenvector of the transform domain host's data autocorrelation matrix as an embedding signature. Correspondingly, it makes the most of the promising signal to interference plus noise ratio (SINR). The authors also proposed in their method that, beneath a (colored) Gaussian assumption on the transform domain host data, the identical resulting signature reduces the host distortion for some marked data retrieval error rate. Above and beyond, it takes full advantage of the Shannon capacity of the covert steganographic link. Besides, such a scheme wherein random-noise-like secret data signals are combined with the host, an inadequate amount of structures cannot continuously distinguish between plain images and their matching stego forms. Employing a superior amount of higher-order statistics structures is able to improve the sensitivity of the structure sensor. Conversely, it increases the computational complexity significantly.

A team of scientists in 2010 suggested, based on code division multiple access (CDMA), a spread spectrum for both the transform domain of an image steganography and the spatial domain in the multimedia messaging service (MMS) (Singh, Khan, Khan, & Singh, 2010). This technique is typically disregarded for the safeguarding of files transmission from a computer to a mobile phone or to another mobile phone, instead by means of a data cable, Bluetooth, infrared, MMS, etc. Employing this CDMA spread spectrum method in the spatial domain offers a stego system of great strength and extra confrontation to threats, related with LSB and DCT methods when employed for steganography in MMS. The outcomes of the experiments revealed that the spread spectrum detection system can offer high robustness to normal signal management, including compression and noise addition. The resultant peak signal-to-noise ratio (PSNR) when approving this method is greater than 50 dB (Singh et al., 2010).

Counting on the SSIS technique, an individual can embed around 0.03 to 0.17 bits per pixel in a 256-level gray-scale image. Worth of note, Kruus et al. (2003) state that robustness can be traded for capability.

Spread spectrum modulation schemes fulfill the maximum data hiding schemes requests, particularly the strength in the contradiction of statistical threats. For this reason, the hidden data are dispersed throughout the image without varying the statistical properties. On the whole, spread spectrum methods can be used in most steganography applications, despite the fact that they are characterized by being a highly mathematical and intricate approach.

2.4.4 Statistical Procedures

Statistical steganographic approaches select statistics that are able to result from an image based on threshold, wherein they obtain a single bit of data. To transmit several bits, an image requests to be fragmented into sub-images, individually consistent to a single bit of the data (Kruus et al., 2003).

Statistical steganographic systems mark the usage of the existence of a "1-bit" steganographic system, wherein one bit of data is embedded in a digital carrier. This procedure is completed by merely adjusting the cover image to an important alteration in the statistical characteristics if "1" is communicated; else, it is left unmoved (Johnson & Katzenbeisser, 2000). A statistical steganography algorithm has been developed by Johnson and Katzenbeisser (2000) from Pitas' watermarking scheme (Pitas, 1996). The latter is like the Patchwork system recommended by Bender, Gruhl, Morimoto, and Lu (1996). With the intention of developing an $l(m)$-bit stego system from a multiple of "1-bit" stego systems, a cover is divided into $l(m)$ disjoint blocks $B_1, B_2, \ldots, B_l(m)$. A secret bit m_i is introduced into the ith block by introducing "1" into B_i if m_i=1. If not, the block is not transformed in the embedding procedure. The discovery of an exact bit is completed through a trial function to differentiate the changed blocks from the unchanged ones, as specified by Johnson and Katzenbeisser (2000):

$$f\left(B_i\right)=\begin{cases}1 \text{ block } B_i \text{ is modified in the embedding process} \\ 0 \quad \text{otherwise}\end{cases} \tag{2.5}$$

Suitably, secret digital watermarks deliver a foundation for a statistical function that is able to be used to encrypt data. Secret digital watermarks are considered to be difficult to sense and difficult to eliminate, while, on the other hand, they are easy to recover given a key. An encoder will watermark those blocks that match a data value of "1". It will leave intact those blocks that match a data value of "0" (Yu-Kuen, Mei-Yi, & Jia-Hong, 2002).

Statistical steganographic approaches, in their simplest practice, are susceptible to cropping, rotating, and scaling attacks, along with any threats that impact against the watermarking technique. To counter these attacks, the sub-images that are simply sub-rectangles of the original image may possibly be designated established on picture elements. For example, the faces in a crowd and an error correction coding could be applied inside the data. These battlements can make the statistical steganographic system roughly as strong as the primary watermarking structure (Kruus et al., 2003).

The central emphasis of this system is established on building an important variation in the statistical features of the cover image if "1" is communicated. Resultantly, this scheme is effortlessly noticed by steganalysis, which includes coding a program that inspects the stego-image arrangement and measures its statistical properties, such as the first-order statistics (histograms) or the second-order statistics (correlations between pixels, distance, direction). Thus, statistical steganographic methods are less promising as compared to other steganographic methods for interconnecting secret data.

2.4.5 Distortion Methods

Distortion methods need to have information about the original cover image during the decoding process. During the this process, the decoder checks the differences between the original and the distorted cover image with the purpose of reinstating the secret data. Alternatively, the encoder enhances a sequence of variations to the cover image. Thus, data are designated as being kept secret by signal distortion (Hamid et al., 2012d,e,f; Radhakrishnan, Shanmugasundaram, & Memon, 2002; Reddy & Raja, 2009). Employing this method, a stego-object is shaped by relating a sequence of alterations to the cover image. This sequence of adjustments is designated to contest the secret data, which is vital in the process of transmission (Katzenbeisser, 2000). The data are encoded at pseudo-randomly selected pixels. If the stego-image is dissimilar from the cover image at the assumed data pixel, then the data bit is "1". If not, the data bit will be "0". The encoder can adjust the "1" value pixels in such a way that the statistical properties of the image are not affected. In turn, this reflects how this method is different from many LSB methods (Johnson & Katzenbeisser, 2000).

The primary methodology in hiding data by this method was text-based. Furthermore, text-based hiding methods are of the distortion form. For instance, the design of a file or the preparation of words might display or reproduce the occurrence of data. In view of one of these systems, individuals can sense the modification of the situations of lines and words where spaces and "invisible" characters have been added to the text, offering a technique for transferring hidden data (Johnson & Katzenbeisser, 2000).

In the context of data embedding methods for a configured text, substantial efforts have been completed by Low and Maxemchuk (1998), Low, Maxemchuk, Brassil, and O'Gorman (1995), and Low, Maxemchuk, and Lapone (1998). In their investigations, text-based steganographic systems are offered in a mode where the space between the successive lines of a text or between the successive words is applied to transfer secret data. It must be noted that, to some extent, the steganographic scheme that manipulates the text format to transfer data can be effortlessly fragmented by retyping the file. Distortion practices can simply be applied to digital images. By means of a comparable method, as the replacement schemes, the source first selects $l(m)$ changed cover-pixels to be used for the data transmission. Such a choice can, once again, be completed by means of PRNGs or pseudo-random permutations. To encode "0" in one pixel, the source leaves the pixel unaffected; to encode "1", a random value is added to the pixel's color.

An image distortion method has been presented by Sandford Ii, Bradley, and Handel (1996), wherein information embedding attempts to adjust the direction of the presence of the redundant information in the cover, instead of to adjust the values themselves. The embedding procedure, hence, upholds a "pair list" (such as a list of pairs of samples whose variance is lower than a particular threshold). The receiver can inverse the embedding procedure if s/he has the right to use to the pair list, like the key in cryptography.

Contrasting several LSB approaches, data hiding by applying distortion methods does not interrupt the somewhat statistical properties of an image. Conversely, the essential task of transferring the cover image limits the advantages of this method. As in some steganographic methods, the cover image must certainly not be used more than once. If the intruder interferes with the stego-image by cropping, rotating, or scaling, the receiver can simply sense the change. In some cases, if the data are encrypted with error-correcting information, the variation can even be reversed and the original data can be recovered in their entirety.

2.4.6 Dossier Embedding Method

Dissimilar image file formats are recognized for taking dissimilar header document structures. Furthermore, using information values, for instance, pixels, palettes, and DCT coefficients, secret data can be hidden in either a header structure or at the end of the document (Yesna, Karen, & Sos, 2007). Perhaps the observation fields in the header of JPEG images typically comprise information hidden by the imperceptible Secrets and Steganozorus. Alternatively, Concealment, JpegX, PGE10, and PGE20 store information at the end of a JPEG image (Cheddad, Condell, Curran, & McKevitt, 2008a). Image storage formats, for instance, tagged image file format (TIFF), GIF, PNG, and Windows metafile (WMF), have a file header that can be employed to hide arbitrary data. In this case, that arbitrary information could be secret data. It is likely able to attach information to several image storage formats without disturbing the image. As soon as the image is managed for display, the image user will decrypt the image size from the file header, and tracking information stored at the end of the document will be overlooked. By means of this system, it is likely to assign a file of some description to a cover image, though the document could be detached from the cover image by merely resaving the image in the same document format (Kruus et al., 2003).

The restrictions of this system are that, even though it has a large payload, it is not problematic to be recognized and overpowered; it is frail when lossy compression and image filtering are taken into account, and that the resaving of an image suggests a comprehensive loss of the hidden information (Cheddad, Condell, Curran, & McKevitt, 2008b).

2.4.7 *Palette Embedding*

In a palette-based image, the point to be noted is that only a subclass of colors from a specific color space is employed to colorize the image. Investigators rely on every single palette-based image format containing two portions. The primary one is a palette that allocates N colors as a list of indexed pairs (i,c_i), where a color vector c_i is allocated to every single index i, and the definite image data identify a palette index for each pixel, instead of identifying the color value itself. The file size becomes reduced through this method when an individual number of color values are employed in the image (Johnson & Katzenbeisser, 2000).

Two of the best general formats of palette-based images are GIF and BMP. On the other hand, due to the disposal of innovative compression methods, their use has weakened (Xuefeng, Then, & Chang-Tsun, 2005). In some scenarios, the palette itself is able to be employed to hide secret data. For the reason that the order of the colors in the palette is not typically of significance, the organization of colors can be employed to hand over data. In principle, hidden data can be embedded by means of the variance between the two colors in the palette (such as one secret data bit for every two colors in the palette). Color palettes are employed to reduce the quantity of data images that are employed to symbolize colors (Kruus et al., 2003; Samaratunge, 2007).

Steganographic data inside the bits of the palette and/or the directories are embedded in the palette-based steganography; one needs to be cautious not to go beyond the determined number of colors (Chih-Hsuan, Zhi-Fang, & Wen-Hsiang, 2004). Using the method of LSB of the palette's color values to hide the secret data is one of the most popular steganographic methods for hiding data in the palette. The reason behind this is that the variations in the LSB do not comprehensively modify the color values and, possibly, will not be noticeable by the naked human eye. The software similar to EzStego implements such a method (Wayner, 2002). In this method, the palette is reordered in such an approach that the adjacent colors in the palette are perceptually alike, a preceding phase before data embedding. This makes the method stronger when the invader tries to rearrange the palette to eliminate the embedded data. On the other hand, palette embedding systems are not very robust, in contrast with other steganography systems. In addition, these methods do not endure simple registering threats, where the invader registers the palette for the determination of abolishing the data without moving the image.

2.4.8 Image Generation Method

Many procedures have been suggested to make the encrypted messages unreadable or as secret as possible. A case in point is the software application of Sam's Big G Play Maker, which hides information by adapting the secret text data into a greater and a somewhat operated text format (EC Council, 2010). The same principle can be employed in image formation, wherein data are transformed to image features and then composed into a comprehensive stego-image. This technique cannot be fragmented by rotating or scaling the image, or by means of lossy compression. Portions of the data will possibly be demolished or vanish due to the procedure of cropping. Nevertheless, it is still likely to retreat the other portions of the data by encrypting the data with error correcting information (England, 1997). Generally, this system practices pseudo-random images. Namely, if a malicious third party spots a cluster of images passing over a system without any aim for the random images present, one could be suspicious that the images comprise secret data and possibly block their transmission (Kruus et al., 2003).

2.4.9 Image Element Adjustment Methods

Approximately, all steganographic methods do not attempt to hide data by means of the definite elements of an image. In its place, they regulate the image elements by entirely unnoticeable means. For instance, they can change the eye color or hair color of an individual in a photo. These adjustments can then be used to transmit the hidden data. It is worth stating that these data will remain persistent in rotations, scaling, and lossy compression. In addition, clipping images will possibly eliminate portions of the embedded message; however, there is still the option for them to be recovered by employing an error correcting coding scheme (Kruus et al., 2003).

The possibility of adjusting items inside images as a method for hiding data has been deliberated by (Bender et al., 2000). It is essential to notice that, when this technique is used, a similar cover image must not be used on more than one occasion. The reason for this is that the elements used will turn out to be superficial. This practice can be accomplished manually with any photo editing software package. Using computer vision systems, such as the procedure of recognizing objects, this process will be more feasible or even automated in the near future (Kruus et al., 2003).

2.4.10 Adaptive Steganography

Adaptive steganography is a distinct case of the spatial and transform techniques. It is presented as "statistics-aware embedding" (Provos & Honeyman, 2003), "masking" (Johnson & Jajodia, 1998), or "model-based" (Sallee, 2004). In adaptive steganography methods, the global statistical characteristics of an image are fundamentally used before any effort is devoted to the agreement with its LSB/ frequency-transformed coefficients. These statistics resolve what variations will possibly be made (Tzschoppe, Baeuml, Huber, & Kaup, 2003). A random adaptive collection of pixels essentially symbolizes this process, depending on the cover image and on the collection of pixels in a block with a great standard deviation. The latter is planned to elude zones of even color; for instance, the smooth areas. This performance marks adaptive steganography, seeking images with current or intentional extra noise and images that establish color complexity (Cheddad et al., 2009a, 2009b, 2010).

The model-based practice, MB1, represented by Sallee (2004), produces a stego-image established on a specified distributed model. A comprehensive Cauchy distribution is implemented to give a lowest distorted stego-image. Inappropriately, this steganographic algorithm can be fragmented by means of the first-order statistics (Böhme & Westfeld, 2004, 2005). Moreover, it can also be noticed by the variance of blocks between a stego-image and its projected form (Yu, Zhao, Ni, & Zhu, 2008).

An adaptive steganographic system for index-based images has been recommended by Chang, Tsai, and Lin (2004). The thinking behind this process is to screen the color palette for color image information hiding. The colors in a palette (code-words) are gathered into sub-clusters as per the connection amongst the code-words. The scope of the sub-cluster is considered to control the hiding capability. In contrast with the conventional LSB process, their trial outcomes display the presentation of the offered system. An enhanced stego-image quality is gained, in contrast with the approaches offered by Fridrich (1999) and Fridrich et al. (2001). An alternative benefit of their system is that the projected system has advanced hiding capabilities.

To disruption the finding of the spatial domain and the frequency domain steganalysis schemes, a genetic algorithm (GA) process was established by Shih (2007) and Yi-Ta and Shih (2006). The stego-image is produced by artificially forging statistical structures with the help of a genetic algorithm. In their investigation, the authors remark that the procedure has to be continual until a predefined situation is reached or a set number of repetitions have been made. The predefined circumstance is the state when the chosen hidden data will possibly be appropriately removed. The foremost disadvantage of the genetic-based algorithms is time complexity, which was not discussed by the authors. Furthermore, it was not specified whether the procedure of computing such a state is done automatically or together with a visual awareness. The recommended GA-based rounding error correction algorithm, although thought-provoking, still suffers from a degree of oversimplification.

A content-based image embedding process, established on segmenting homogenous gray-scale areas by means of a watershed system together with fuzzy c-means (FCM)m was offered by Jun, Hongru, Xiaolu, and Zhi (2009). In this technique, the secret information is encoded by a chaotic map before being embedded. Then and there, the cover image is segmented by a watershed algorithm and FCM. Afterwards, the features of an individual section are removed and the secret information is embedded into the cover image, conferring to the outcome of the features removal. The entropy is then designed for the individual section. The attained entropy values direct the embedding power. The four LSBs of each of the cover's RGB primaries are used if the entropy surpasses a defined threshold; else, merely two LSBs for each are used (Cheddad, 2009). The fragile fact of this system lies in its sensitivity to force deviations, a problem that strictly marks the removal of the correct secret bits. In the work of Jun et al. (2009), the investigators likewise testified the usage of a logistic map to encode the secret bit stream which appears susceptible to a chosen-plaintext attack (CPA).

An innovative steganographic method was offered by Cheddad (2009), where a robust steganographic scheme named Steganoflage was recognized. Steganoflage follows an object-oriented method wherein skin-tone perceived zones in the image are selected for embedding, wherever conceivable. The approved embedding technique is the reflected binary gray code (RBGC) in the wavelet domain. The attained outcomes are encouraging and are used to classify the provision and stability of the established algorithm. A sequence of stimulating uses is publicized; for instance, opposing digital counterfeit, multilayer security for patients' data storage, transmission and digital restoration of missing signals, etc. In their system, the foremost restraint that needs to be taken into account is selecting embedding in skin-tone areas. This restraint disturbs somewhat the liberty of choosing the cover image. Instead, skin-tone embedding bounds the hiding capability. At that point, there is no strong argument regarding the attained payload capability. Lastly, the authors adopt to encode the image itself, which will possibly entice the responsiveness of observers in the direction of deliberate or unintended threats.

The adaptive more surrounding pixels using (A-MSPU) method, which recovers the imperceptibility difficulties of the multiple base notational system (MBNS), has been suggested by Afrakhteh and Ibrahim (2010). This method emphasizes responsiveness to the procedure of embedding in the edge zones of a cover image, while re-expressing the secret bits in numerous base notational schemes. The submitted method uses a similar probability parameter to acquire the dispersed secret bits. Moreover, it tests nearby pixels by a determined amount to decide the capability of every single marked pixel. Most steganographic methods examine either three or four adjacent pixels of a marked pixel. The authors' suggested method is capable of utilizing all eight adjacent neighbors, an asset that advances the imperceptibility value. It is noteworthy that the main focus of this research revolves around information hiding using adaptive steganography in the frequency domain, such as DCT and DWT. These techniques are not too prone to attacks, especially when the hidden message is small. This is because they are able to alter coefficients in the transform domain; thus, image distortion is kept to a minimum.

2.5 Performance Capacity

As a performance measure for image distortion due to the process of embedding, the well-known PSNR, which is categorized under difference distortion metrics, can be applied to stego-images (Cheddad et al., 2010). It is defined as:

$$\text{PSNR} = 10\ \log\left(\frac{C^2_{\max}}{\text{MSE}}\right) \tag{2.6}$$

where MSE denotes the mean square error, which is given as:

$$\text{MSE} = \frac{1}{MN}\sum\nolimits_{x=1}^{M}\sum\nolimits_{Y=1}^{N}\left(S_{xy} - C_{xy}\right)^2 \tag{2.7}$$

Here, $C_{\max}$ indicates the maximum value in the image, for example:

$$C_{\max} \le \begin{cases} 1 & \text{in double precision images} \\ 255 & \text{in 8-bit unsigned integer intensity images} \end{cases} \tag{2.8}$$

In addition, x and y represent the image coordinates, while M and N denote the dimensions of the image, S_{xy} is the resultant stego-image, and C_{xy} stands for the cover image. In the above, $C_{\max}$ is fixed to 255, to be measured as a default value for 8-bit images (Drew & Bergner, 2008; Hashad, Madani, & Wahdan, 2005; Kermani & Jamzad, 2005; Li & Wang, 2007; Yu, Chang, & Lin, 2007). It can also be an image that has merely up to 253 gray colors. It must be noticed that, when $C_{\max}$ is elevated to the power of 2, it results in a severe variation in the PSNR value. Therefore, $C_{\max}$ is measured as the definite concentrated value relatively, rather than the major likely value. The PSNR is frequently communicated on a logarithmic scale in decibels (dB). PSNR values under 30 dB specify a poor value (such as when the distortion triggered by embedding is vibrant). On the other hand, the high-quality stego-image must achieve a PSNR of 40 dB or more (Hernandez-Castro, Blasco-Lopez, Estevez-Tapiador, & Ribagorda-Garnacho, 2006; Zamani, Manaf, & Abdullah, 2012).

2.6 Steganalysis

The goal of steganography is to avoid drawing attention to the transmission of a secret message (Hamid et al., 2013a,b). If the communication is doubtful, then this objective is not met. Steganalysis is the art and science that aims to preserve the hidden data inside a cover dossier. An examiner, identified as a steganalyst, carries out steganalysis on a digital folder in an effort to discover and remove the stego data (Chen, 2010). At first, the science of steganalysis is intended to perceive or approximate the occurrence of the hidden data established on information handover, without making any assumptions on the steganography algorithm. In digital image steganalysis, an analyst has three goals. One has to decide if embedded data exist, then use an embedding technique to create the stego-image, and, lastly, remove the hidden data (Rodriguez, 2008).

Steganalysis is accomplished using diverse image processing methods, such as image filtering, rotating, cropping, and translating. In addition, it can be completed by implementing a coding program to check the stego-image organization and measure its statistical properties. Statistical properties dimensions contain first-order statistics (histograms) or second-order statistics (correlation between pixels, direction, and distance). Besides, JPEG double compression and the distribution of DCT coefficients can be measured as a strong indication for the practice of DCT-based image steganography (Cheddad et al., 2010).

In 2006, Kharrazi, in his report, specified that one must recognize that steganalysis algorithms are considered fruitful if they are capable of noticing the presence of data, in place of decrypting the data itself (Kharrazi, 2006). The concluding procedure would be very difficult if the data are encoded with one of the robust cryptography algorithms. Nonetheless, in the current scenario, there are distinct approaches which can estimate the size of the embedded data and perceive the occurrence of the secret data. Johnson and Jajodia (1998) bring together subsequent descriptions, which are implemented by the steganalysis community:

- Stego-only attack: The stego dossier is the single piece offered for investigation;
- Known cover attack: The novel cover and stego dossier are equally accessible for investigation;
- Known message attack: Sometimes, the invader will possibly recognize the hidden data. Nevertheless, evaluating the stego-image for patterns that correspond to the hidden data will possibly be helpful to break out the structure in the forthcoming steps. Even with the data, this will possibly be very tough and might even be reflected akin to the stego-only threat;
- Chosen stego attack: The stego dossier and the tool used (algorithm) are equally recognized;
- Chosen stego message attack: The steganalyst produces stego documents from a known steganography tool using a chosen stego message. The idea behind this threat is to deduce the matching patterns in the stego-image that will possibly direct to the usage of specific steganography methods or algorithms; and
- Known stego attack: The cover file, stego dossier, and stego means are identified.

2.7 Evaluation of Previously Declared Methods

Existing algorithms, in regard to image steganography, are not immune to frailty and robustness issues. Accordingly, it is imperative to deduce the most appropriate method in order to be functional. As described earlier, there are three key constraints

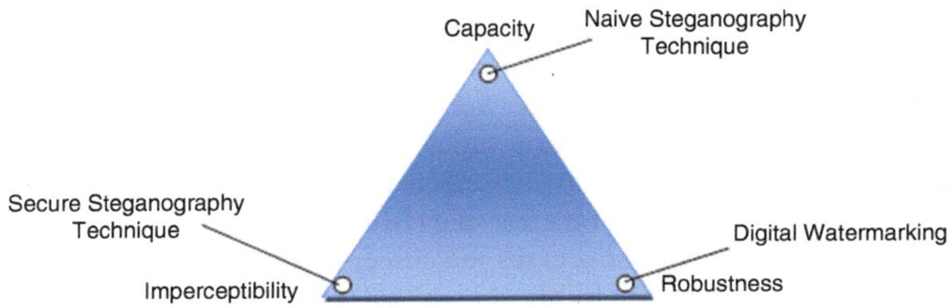

Fig. 2.8 The competing factors in steganographic systems. (Fridrich, 1999)

that are used to size the presentation of the steganographic structure (Amirtharajan & Rayappan, 2012). Figure 2.8 displays the affiliation between steganography parameters (Fridrich, 1999). These parameters include undetectability (imperceptibility), strength, and payload capability. The subsequent sub-sections relate the previously discussed steganographic systems to the three opposing parameters:

- The LSB system in the spatial domain is a useful means to cover data. Conversely, it is susceptible to minor variations resulting from image processing or lossy compression (Johnson & Jajodia, 1998). Though LSB methods can hide huge amounts of data, with an extraordinary payload capability, they often reimburse the statistical properties of the image and, hence, specify a small strength contrary to statistical threats in addition to image handling.
- The encouraging methods DCT, DWT, and adaptive steganography are not immune to threats; specifically, when the hidden data are minor. This can be vindicated in agreement to the approach that they alter the coefficients in the transform domain. Largely, these systems are inclined to have a poorer payload in comparison to the spatial domain algorithms (Cheddad et al., 2010). The experiments on the DCT coefficients introduced some promising results that diverted investigators' efforts in the direction of JPEG images. Functioning at a level similar to that of DCT, steganography gains greater control and becomes less susceptible to statistical threats. Embedding in the DWT domain discloses a sort of productive outcome and its embedding performance surpasses that of DCT as far as compression survival is concerned (Cheddad, 2009).
- Spread spectrum practices are commonly pretty robust in countering statistical threats, since the hidden data are spread all over the image. Spread spectrum encrypting is widely used in military communications by reason of its strength contrary to recognition. When data are embedded, an invader cannot simply document it and it will be hard to remove without knowing the appropriate keys. SISS is very valuable for steganography for the reason of extraordinary capability and the great difficulty offered in the procedures of discovery and removal. However, this scheme is still susceptible to obliteration by compression and image processing. A determined invader can rather effortlessly give and take the

embedded information by means of some digital processing, for instance, noise reduction filters, used in decoding to approximate the original cover. Furthermore, the simple compromise in SSIS is between the error rate and the quantity of data to be embedded. The latter, in return, varies with the variation of the power of the added noise. The used ECC must be selected cautiously so as to retain a small power without increasing the bit error rate (BER).

- The statistical systems in most cases are susceptible to cropping, rotating, and scaling attacks, accompanied by any threats that work contrary to the watermarking method. Resistances could be measured to make the statistical methods as strong as the watermarking system, as explained in detail in Sect. 2.4.4. The payload capability and invisibility are subject to the cover image selection.
- In several disparate LSB systems, distortion practices do not reduce some statistical properties of the image. In contrast, the necessity to transmit the cover image over a secure channel confines the value of this practice. As in any steganographic method, the cover image must not be used more than once. If an invader modifies the stego-image by cropping, rotating, or scaling, the modification can simply be assumed by the receiver and just reversed to the correct orientation, where the message encrypted with error correcting data can be fully recovered. Error correcting data also support if the stego-image is filtered over a lossy compression structure, for instance JPEG. Implementing this practice limits the hidden data capability, while adding distortion to the cover image is the foundation of embedding algorithms. As a consequence, the distorted image will be extra susceptible to the HVS.
- Systems that adapt image file formatting data have a huge payload. Nonetheless, they have the resulting disadvantages: they are effortlessly perceived and overpowered; they are not strong contrary to lossy compression and image filters; and the matter of saving the image one more time absolutely eradicates the hidden information (Cheddad et al., 2008b).
- Hiding data via steganographic methods that adjust the elements in the visual image results in a stego-image that will continue rotating, scaling, and enduring much lossy compression, like JPEG. A rational payload capability can be accomplished with this practice too. Table 2.1 summarizes the assessment of the systems described throughout this chapter.

Table 2.1 A comparison of image steganography methods

Steganography parameter	LSB	Transform domain	Spread spectrum	Statistical techniques	Distortion techniques	File and palette embedding
Imperceptibility	High[a]	High	High	Medium[a]	Low	High[a]
Robustness	Low	High	Medium	Low	Low	Low
Payload capacity	High	Low	High	Low[a]	Low	High

[a]Specifies dependency on the used cover image

References

Abdelwahab, A. A., & Hassaan, L. A. (2008, 18–20 March 2008). *A discrete wavelet transform based technique for image data hiding*. Paper presented at the radio science conference, 2008. NRSC 2008. National.

Abdul-mahdi, N. H., Yahya, A., Ahmad, R. B., & Al-Qershi, O. M. (2013). Secured and robust information hiding scheme. *Procedia Engineering Journal, 53*, 463–471.

Afrakhteh, M., & Ibrahim, S. (2010, 25–27 June 2010). *Adaptive steganography scheme using more surrounding pixels*. Paper presented at the Internatioal Conference on Computer Design and Applications (ICCDA), 2010.

Alturki, F., & Mersereau, R. (2001, 7–10 Oct 2001). *Secure blind image steganographic technique using discrete Fourier transformation*. Paper presented at the International Conference on Image Processing, 2001. Proceedings.

Amirtharajan, R., & Rayappan, J. B. B. (2012). Inverted pattern in inverted time domain for icon steganography. *Information Technology Journal, 11*(5), 587–595.

Anand, D., & Niranjan, U. C. (1998, 29 Oct–1 Nov 1998). *Watermarking medical images with patient information*. Paper presented at the Engineering in Medicine and Biology Society, 1998. Proceedings of the 20th Annual International Conference of the IEEE.

Anderson, R. J., & Petitcolas, F. A. P. (2006). On the limits of steganography. *IEEE Journal on Selected Areas in Communications, 16*(4), 474–481.

Areepongsa, S., Kaewkamnerd, N., Syed, Y. F., & Rao, K. R. (2000). *Exploring steganography for low bit rate wavelet based coder in image retrieval system*. Paper presented at the TENCON 2000. Proceedings.

Areepongsa, S., Syed, Y. F., Kaewkamnerd, N., & Rao, K. R. (2000). *Steganography for a low bit-rate wavelet based image coder*. Paper presented at the International Conference on Image Processing, 2000. Proceedings.

Bender, W., Gruhl, D., Morimoto, N., & Lu, A. (1996). Techniques for data hiding. *IBM Systems Journal, 35*(3.4), 313–336.

Bender, W., Butera, W., Gruhl, D., Hwang, R., Paiz, F. J., & Pogreb, S. (2000). Applications for data hiding. *IBM Systems Journal, 39*(3–4), 547–568.

Böhme, R., & Westfeld, A. (2004). Breaking Cauchy model-based JPEG steganography with first order statistics. In P. Samarati, P. Ryan, D. Gollmann, & R. Molva (Eds.), *Computer security – ESORICS 2004* (Vol. 3193, pp. 125–140). Berlin/Heidelberg: Springer.

Böhme, R., & Westfeld, A. (2005). *Exploiting preserved statistics for steganalysis* Springer Science & Business Media. ISBN 364214313X, 9783642143137.

Chandramouli, R., Kharrazi, M., & Memon, N. (2004). *Image steganography and steganalysis: Concepts and practice Lecture notes in computer science* (Vol. 2939). Berlin, Heidelberg: Springer.

Chang, C.-C., & Tseng, H.-W. (2004). A steganographic method for digital images using side match. *Pattern Recognition Letters, 25*(12), 1431–1437.

Chang, C.-C., Tsai, P., & Lin, M.-H. (2004). *An adaptive steganography for index-based images using codeword grouping*. Paper presented at the Proceedings of the 5th Pacific Rim Conference on Advances in Multimedia Information Processing – Volume Part III, Tokyo, Japan.

Cheddad, A. (2009). *Steganoflage: A new image steganography algorithm*. Doctor of Philosophy Thesis, University of Ulster, Northern Ireland, UK.

Cheddad, A., Condell, J., Curran, K., & McKevitt, P. (2008a, March 31–April 4). *Biometric Inspired Digital Image Steganography*. Paper presented at 15th Annual IEEE International Conference and Workshop on the Engineering of Computer Based Systems (ECBS), 2008.

Cheddad, A., Condell, J., Curran, K., & McKevitt, P. (2008b, 28–30 May). *Enhancing Steganography in Digital Images*. Paper presented at the Canadian Conference on Computer and Robot Vision, 2008. CRV '08.

Cheddad, A., Condell, J., Curran, K., & McKevitt, P. M. (2009a). A secure and improved self-embedding algorithm to combat digital document forgery. *Signal Processing, 89*(12), 2324–2332.

Cheddad, A., Condell, J., Curran, K., & McKevitt, P. (2009b). A skin tone detection algorithm for an adaptive approach to steganography. *Signal Processing, 89*(12), 2465–2478.

Cheddad, A., Condell, J., Curran, K., & Kevitt, P. M. (2010). Digital image steganography: Survey and analysis of current methods. *Signal Processing, 90*(3), 727–752.

Chen, M.-C. (2010). *Image security and recognition system.* Ph.D. dissertation, the University of Texas at San Antonio, Texas, United States.

Chen, W.-Y. (2007). Color image steganography scheme using set partitioning in hierarchical trees coding, digital Fourier transform and adaptive phase modulation. *Applied Mathematics and Computation, 185*(1), 432–448.

Chih-Hsuan, T., Zhi-Fang, Y., & Wen-Hsiang, T. (2004). Adaptive data hiding in palette images by color ordering and mapping with security protection. *IEEE Transactions on Communications, 52*(5), 791–800.

Drew, M. S., & Bergner, S. (2008). Spatio-chromatic decorrelation for color image compression. *Image Communication, 23*(8), 599–609.

EC Council. (2010). *Attack phases. EC Council Press.* Clifton Park, NY: Course Technology/ Cengage Learning.

El-Emam, N. N. (2007). Hiding a large amount of data with high security using steganography algorithm. *Journal of Computer Science, 3*(4), 223–232.

England, N. (1997). New image generation techniques. *Computer Graphics and Applications, IEEE, 17*(1), 39–39.

Farid, H. (2009). Image forgery detection. *Signal Processing Magazine IEEE, 26*(2), 16–25.

Fridrich, J. (1999). Applications of data hiding in digital images. *Proceedings of the Fifth International Symposium on Signal Processing and its Applications*, 1999, ISSPA '99. Australia.

Fridrich, J., & Goljan, M. (2002). Practical steganalysis of digital images - state of the art. Electronic Imaging, 19–25 January 2002, San Jose, California, United States.

Fridrich, J., Goljan, M., & Hogea, D. (2003). *Steganalysis of JPEG images: Breaking the F5 algorithm lecture notes in computer science* (Vol. 2578). Berlin, Heidelberg: Springer.

Fridrich, J., Goljan, M., & Rui, D. (2001). Detecting LSB steganography in color, and gray-scale images. *MultiMedia, IEEE, 8*(4), 22–28.

Geetha, S., Kabilan, V., Chockalingam, S. P., & Kamaraj, N. (2011). Varying radix numeral system based adaptive image steganography. *Information Processing Letters, 111*(16), 792–797.

Gkizeli, M., Pados, D. A., & Medley, M. J. (2007). Optimal signature design for spread-spectrum steganography. *IEEE Transactions on Image Processing, 16*(2), 391–405.

Hamid, N., Yahya, A., & Ahmad, R. B. (2011a). *Performance analysis of image steganography techniques*. Paper presented at AKEPT's 1st annual young researchers conference (AYRC X32011), Kuala Lumpur, Malaysia.

Hamid, N., Yahya, A., & Ahmad, R. B. (2011b, 22–23 October). *An overview of image steganography techniques*. Paper presented at the International Postgraduate Conference on Engineering (IPCE2011), Perlis, Malaysia.

Hamid, N., Yahya, A., Ahmad, R. B., & Al-Qershi, O. M. (2012a, 10–12 April). Characteristic region based image steganography using speeded-up robust features technique. Paper presented at the 1st IEEE International Conference on Future Communication Network (ICFCN'12). Iraq, Baghdad.

Hamid, N., Yahya, A., Ahmad, R. B., & Al-Qershi, O. M. (2012b). Secured and Robust Information Hiding Scheme. Paper presented at the Malaysian Technical Universities Conference on Engineering and Technology (MUCET2012), Kangar, Perlis, Malaysia.

Hamid, N., Yahya, A., Ahmad, R. B., & Al-Qershi, O. M. (2012c, 18–20 June). *Blind image steganography scheme using speeded-up robust features technique*. Paper presented at the 2nd International Malaysia-Ireland joint symposium on engineering, science and business (IMiEJS2012), Kuala Lumpur, Malaysia.

Hamid, N., Yahya, A., Ahmad, R. B., & Al-Qershi, O. M. (2012d). An improved robust and secured image Steganographic scheme. *International Journal of Electronics and Communication Engineering & Technology (IJECET), 3*(2), 484–496.

Hamid, N., Yahya, A., Ahmad, R. B., & Al-Qershi, O. M. (2012e). A comparison between using SIFT and SURF for characteristic region based image steganography. *International Journal of Computer Science Issues (IJCSI), 9*(3), 110–116.

Hamid, N., Yahya, A., Ahmad, R. B., & Al-Qershi, O. M. (2012f). Image steganography techniques: An overview. *International Journal of Computer Science and Security (IJCSS), 6*(3), 168–178.

Hamid, N., Yahya, A., Ahmad, R. B., Najim, D., & Kanaan, L. (2013a). Steganography in image files: A survey. *Australian Journal of Basic and Applied Sciences., 7*(1), 35–55.

Hamid, N., Yahya, A., Ahmad, R. B., Najim, D., & Kanaan, L. (2013b). Enhancing the robustness of digital image steganography using ECC and redundancy. *WULFENIA Journal, 20*(4), 153–169.

Hashad, A. I., Madani, A. S., & Wahdan, A. E. M. A. (2005, 5–6 Dec 2005). *A robust steganography technique using discrete cosine transform insertion.* Paper presented at the ITI 3rd International Conference on Information and Communications Technology, 2005. Enabling technologies for the new knowledge society.

Hernandez-Castro, J. C., Blasco-Lopez, I., Estevez-Tapiador, J. M., & Ribagorda-Garnacho, A. (2006). Steganography in games: A general methodology and its application to the game of Go. *Computers and Security, 25*(1), 64–71.

Information Hiding. In J. Fridrich (Ed.), (Vol. 3200, pp. 359–379). Berlin/Heidelberg: Springer.

Jamil, T. (1999). Steganography: The art of hiding information in plain sight. *Potentials IEEE, 18*(1), 10–12.

Johnson, N. F. (1995). Steganography. Technical report. Retrieved August 24, 2012.

Johnson, N. F. (2009, 2011). Steganography software. Retrieved August 26, 2012.

Johnson, N. F., & Katzenbeisser, S. (2000). A survey of steganographic techniques. In S. Katzenbeisser & F. A. P. Petitcolas (Eds.), *Information hiding techniques for steganography and digital watermarking* (pp. 43–78). London: Artech House.

Johnson, N., & Jajodia, S. (1998). *Steganalysis of images created using current steganography software*

Judge, J. C. (2001). Steganography: Past, present, future. Retrieved August 24, 2012

Jun, K., Hongru, J., Xiaolu, L., & Zhi, Q. (2009, 22–24 Jan). *A novel content-based information hiding scheme.* Paper presented at theInternational Conference on Computer Engineering and Technology, 2009. ICCET '09.

Juneja, M., & Sandhu, P. S. (2009, 27–28 Oct). *Designing of robust image steganography technique based on LSB insertion and encryption.*Paper presented at theInternational Conference on Advances in Recent Technologies in Communication and Computing, 2009. ARTCom '09.

Katzenbeisser, S. C. (2000). Principles of steganography. In S. Katzenbeisser & F. A. P. Petitcolas (Eds.), *Information hiding techniques for steganography and digital watermarking* (pp. 17–41). Boston/London: Artech House.

Kekre, H. B., Mishra, D., Shah, S., Shah, R., & Thakkar, C. (2012). Row-wise DCT plane sectorization in CBIR. *International Journal of Computer Applications, 46*(4), 29–35.

Kermani, Z. Z., & Jamzad, M. (2005, December 21-21). *A robust steganography algorithm based on texture similarity using Gabor filter.* Paper presented at the Fifth IEEE International Symposium on Signal Processing and Information Technology, 2005. Proceedings.

Kharrazi, M. (2006). *Image steganography and steganalysis.* PhD Dissertation, Polytechnic University, (30250643).

Ki-Hyun, J., Kyeoung-Ju, H., & Kee-Young, Y. (2008, August 28–30). *Image Data Hiding Method Based on Multi-Pixel Differencing and LSB Substitution Methods.* Paper presented at theInternational Conference on Convergence and Hybrid Information Technology, 2008. ICHIT '08.

Kruus, P., Scace, C., Heyman, M., & Mundy, M. (2003). A survey of steganographic techniques for image files. *Advanced Security Research Journal, V(I)*, 41–52.

Li, X., & Wang, J. (2007). A steganographic method based upon JPEG and particle swarm optimization algorithm. *Information Science., 177*(15), 3099–3109.

Li, B., He, J., Huang, J., & Shi, Y. Q. (2011). A survey on image steganography and steganalysis. *Journal of Information Hiding and Multimedia Signal Processing, 2*(2), 142–172.

Liao, X., Wen, Q.-y., & Zhang, J. (2011). A steganographic method for digital images with four-pixel differencing and modified LSB substitution. *Journal of Visual Communication and Image Representation, 22*(1), 1–8.

Lin, C.-C., & Shiu, P.-F. (2010). High capacity data hiding scheme for DCT-based images. *Journal of Information Hiding and Multimedia Signal Processing, 1*(3), 220–240.

Lin, Y.-K. (2012). High capacity reversible data hiding scheme based upon discrete cosine transformation. *Journal of Systems and Software, 85*(10), 2395–2404. https://doi.org/10.1016/j.jss.2012.05.032

Low, S. H., Maxemchuk, N. F., Brassil, J. T., & O'Gorman, L. (1995, 2–6 Apr 1995). *Document marking and identification using both line and word shifting*. Paper presented at the INFOCOM '95. Fourteenth Annual Joint Conference of the IEEE Computer and Communications Societies. Boston, Massachusetts.

Low, S. H., & Maxemchuk, N. F. (1998). Performance comparison of two text marking methods. *IEEE Journal on Selected Areas in Communications, 16*(4), 561–572.

Low, S. H., Maxemchuk, N. F., & Lapone, A. M. (1998). Document identification for copyright protection using centroid detection. *IEEE Transactions on Communications, 46*(3), 372–383.

Lu, C.-S. (2005). *Multimedia security: Steganography and digital watermarking techniques for protection of intellectual property*. Hershey, Pennsylvania: Idea Group Publishing.

Marvel, L. M., Boncelet Jr., C. G., & Retter, C. T. (1999). Spread spectrum image steganography. *IEEE Transactions on Image Processing, 8*(8), 1075–1083.

Marvel, L. M., Retter, C. T., & Boncelet, C. G., Jr. (1998a, 4–7 Oct). *Hiding information in images*. Paper presented at the International Conference on Image Processing, 1998. ICIP 98. Proceedings.

Marvel, L. M., Retter, C. T., & Boncelet, C. G., Jr. (1998b, 18–21 Oct). *A methodology for data hiding using images*. Paper presented at the Military Communications Conference, 1998. MILCOM98. Proceedings., IEEE.

Morkel, T., Eloff, J. H. P., & Olivier, M. S. (2005). *An overview of image steganography.*Paper presented at the Proceedings of the Fifth Annual Information Security South Africa Conference (ISSA2005), Sandton, South Africa.

Noda, H., Niimi, M., & Kawaguchi, E. (2006). High-performance JPEG steganography using quantization index modulation in DCT domain. *Pattern Recognition Letters, 27*(5), 455–461.

Park, Y.-R., Kang, H.-H., Shin, S.-U., & Kwon, K.-R. (2005). *A steganographic scheme in digital images using information of neighboring pixels. Advances in natural computation. ICNC 2005. Lecture notes in computer science* (Vol. 3612). Berlin, Heidelberg: Springer.

Paulson, L. D. (2006). News briefs. *Computer, 39*(8), 25–27.

Petitcolas, F. A. P. (2000). Introduction to information hiding. In S. Katzenbeisser & F. A. P. Petitcolas (Eds.), *Information hiding techniques for steganography and digital watermarking* (pp. 1–12). Boston/London: Artech House.

Pickholtz, R., Schilling, D., & Milstein, L. (1982). Theory of spread-spectrum communications—atutorial. *IEEE Transactions on Communications, 30*(5), 855–884.

Pitas, I. (1996, 16–19 Sep). *A method for signature casting on digital images*. Paper presented at the International Conference on Image Processing, 1996. Proceedings.

Potdar, V. M., Han, S., & Chang, E. (2005, 10–12 Aug). *A survey of digital image watermarking techniques*. Paper presented at the 3rd IEEE International Conference on Industrial Informatics, 2005. INDIN '05.

Prasad, M. S., Janeyulu, S. N., Krishna, C. G., & Nagaraju, C. (2009). A novel information hiding technique for security by using image steganography. *Journal of Theoretical and Applied Information Technology, 8*(1), 35–39.

Provos, N. (2001). *Defending against statistical steganalysis*. Paper presented at the Proceedings of the 10th conference on USENIX Security Symposium – Volume 10, Washington, D.C.

Provos, N., & Honeyman, P. (2003). Hide and seek: An introduction to steganography. *IEEE Security & Privacy, 1*(3), 32–44.

Radhakrishnan, R., Shanmugasundaram, K., & Memon, N. (2002, 9–11 Dec). *Data masking: a secure-covert channel paradigm.* Paper presented at the IEEE Workshop on Multimedia Signal Processing, 2002.

Reddy, H. S. M., & Raja, K. B. (2009). High capacity and security steganography using discrete wavelet transform. *International Journal of Computer Science and Security (IJCSS), 3*(6), 462–472.

Rodriguez, B. M., II. (2008). *Multi-class classification for identifying JPEG steganography embedding methods.* Ph.D. dissertation, Air Force Institute of Technology, Ohio, United States.

Sallee, P. (2004). Model-Based Steganography Digital Watermarking. Lecture Notes in Computer Science, vol 2939. Springer, Berlin, Heidelberg.

Samaratunge, S. G. K. D. N. (2007). *New steganography technique for palette based images International Conference on Industrial and Information Systems*, 2007. ICIIS 2007, Sri Lanka

Sandford Ii, M. T., Bradley, J. N., & Handel, T. G. (1996). Data embedding method. 226–259.

Shaou-Gang, M., Chin-Ming, H., Yuh-Show, T., & Hui-Mei, C. (2000). *A secure data hiding technique with heterogeneous data-combining capability for electronic patient records.* Paper presented at the Engineering in Medicine and Biology Society, 2000. Proceedings of the 22nd Annual International Conference of the IEEE.

Shejul, A. A., & Kulkarni, U. L. (2011). A secure skin tone based steganography using wavelet transform. *International Journal of Computer Theory and Engineering, 3*(1), 16–22.

Shen, C., Zhang, H., Feng, D., Cao, Z., & Huang, J. (2007). Survey of information security. *Science in China Series F: Information Sciences, 50*(3), 273–298.

Shih, F. Y. (2007). *Digital watermarking and steganography: Fundamentals and techniques.* Boca Raton, FL: Taylor & Francis.

Singh, R. P., Khan, M. A. A., Khan, M., & Singh, N. (2010). Spread spectrum image steganography in multimedia messaging service of mobile phones. *International Journal of Electronics Engineering, 2*(2), 365–369.

Smith, J. R., & Comiskey, B. O. (1996). *Modulation and information hiding in images.* Paper presented at the Proceedings of the First International Workshop on Information Hiding.

Solanki, K., Sarkar, A., & Manjunath, B. S. (2007). *YASS: Yet another steganographic scheme that resists blind steganalysis.* Paper presented at the Proceedings of the 9th international conference on Information hiding, Saint Malo, France.

Syed, Y. F. (1999). *A low bit rate wavelet-based image coder for transmission over hybrid networks.* The University of Texas at Arlington.

Syed, Y. F., & Rao, K. R. (1999, 24–27 Oct). *Scalable low bit rate coding using an HC-RIOT coder.* Paper presented at the Thirty-Third Asilomar Conference on Signals, Systems, and Computers, 1999. Conference Record.

Tzschoppe, R., Baeuml, R., Huber, J., & Kaup, A. (2003). *Steganographic system based on higher-order statistics.* Paper presented at the Security and Watermarking of Multimedia Contents V, Santa Clara, California, USA.

Wang, C.-M., Wu, N.-I., Tsai, C.-S., & Hwang, M.-S. (2008). A high quality steganographic method with pixel-value differencing and modulus function. *Journal of Systems and Software, 81*(1), 150–158.

Wayner, P. (2002). *Disappearing cryptography information hiding: Steganography & watermarking.* San Francisco: Morgan Kaufmann Publishers Inc ©2002 ISBN:1558607692.

Westfeld, A. (2001). *F5—a steganographic algorithm: High capacity despite better steganalysis.* Paper presented at the 4th International Workshop on Information Hiding Lecture Notes in Computer Science, Vol. 2137. Springer, Berlin, Heidelberg.

Westfeld, A., & Pfitzmann, A. (2000). Attacks on Steganographic systems. *Lecture notes in computer science, Vol 1768.* Berlin, Heidelber: Springer.

Widadi, K. C., Ainianta, P. H., & Chan Choong, W. (2005). *Blind steganography using direct sequence/frequency hopping spread spectrum technique.* Paper presented at the Fifth International Conference on Information, Communications and Signal Processing, 2005 Bangkok, Thailand.

Wu, D.-C., & Tsai, W.-H. (2003). A steganographic method for images by pixel-value differencing. *Pattern Recognition Letters, 24*(9–10), 1613–1626.

Wu, H. C., Wu, N. I., Tsai, C. S., & Hwang, M. S. (2005). Image steganographic scheme based on pixel-value differencing and LSB replacement methods. *IEE Proceedings - Vision, Image and Signal Processing, 152*(5), 611–615.

Xuefeng, W., Then, Y., & Chang-Tsun, L. (2005, 11–14 Sept). *A palette-based image steganographic method using colour quantisation.* Paper presented at the IEEE International Conference onimage processing, 2005. ICIP 2005 Genova, Italy.

Yang, C.-H., & Wang, S.-J. (2006). *A steganographic method for digital images by multi-pixel differencing.* Paper presented at the International Computer Symposium, Taipei, Taiwan.

Yesna, Y., Karen, P., & Sos, A. (2007). *New quantization matrices for JPEG steganography.* Paper presented at the Mobile Multimedia/Image Processing for Military and Security Applications 2007, USA.

Yi-Ta, W., & Shih, F. Y. (2006). Genetic algorithm based methodology for breaking the steganalytic systems. *IEEE Transactions on Systems, Man, and Cybernetics, Part B: Cybernetics, 36*(1), 24–31.

Yu, L., Zhao, Y., Ni, R., & Zhu, Z. (2008). PM1 steganography in JPEG images using genetic algorithm. *Soft Computing, 13*(4), 393–400.

Yu, Y.-H., Chang, C.-C., & Lin, I.-C. (2007). A new steganographic method for color and grayscale image hiding. *Computer Vision and Image Understanding, 107*(3), 183–194.

Yue, L., Chang-Tsun, L., & Chia-Hung, W. (2007, 29–31 Aug). *Protection of mammograms using blind steganography and watermarking.* Paper presented at the Third international symposium on Information Assurance and Security, 2007. IAS 2007 Manchester, UK.

Yu-Kuen, H., Mei-Yi, W., & Jia-Hong, L. (2002). *Hierarchic texture classification using statistical steganography techniques.* Paper presented at the 14th International Conference on Digital Signal Processing, 2002. DSP 2002 Santorini, Greece, Greece.

Zamani, M., Manaf, A. B. A., & Abdullah, S. M. (2012). *Correlation between PSNR and size ratio in audio steganography.* Paper presented at the 11th International Conference on Telecommunications and Informatics. *Proceedings of the 11th International Conference on Signal Processing*, Saint Malo, Mont Saint-Michel, France.

Chapter 3
Characteristic Region-Based Image Steganography

Abstract For most of the current steganography techniques, the information-hiding process modifies almost all cover components. Hiding data in the whole image may affect visual quality and increases the possibility of data loss after any possible attacks. In this chapter, a new region-based steganography method, CR-BIS, which hides data in the robust regions of the image, is proposed. First, the secret data are encrypted via a highly secure encryption algorithm. Second, SURF is used to locate the strongest sections in the image. Then data embedding is accomplished in a content-based style by varying the wavelet transform coefficients of those strong sections. The robustness of the proposed algorithm increases when second-level DWT is used to hide data, especially against JPEG compression. However, applying the same scheme to the median and the low-pass filters remains difficult. Utilizing higher DWT levels is useful to enhance the robustness.

3.1 Introduction

This section debates the main structure of the proposed scheme and examines the theoretical aspects of the CR-BIS in detail. The chapter also explains the suggested framework which links three components, namely encryption, characteristic region detection, and steganography. In this chapter, the concept of characteristic region embedding is introduced into information-hiding in general and mainly to steganography. The characteristic regions in the image, rather than the whole image or certain locations in the image, are used to conceal secret information. Such a method will increase the security of the embedding process. To detect the characteristic regions, SURF technique is employed in the proposed algorithm. Those characteristic regions are described as robust to various transformations, such as rotation, scaling, cropping, and affine transformations, which enables them to resist possible attacks. This condition implies that those regions can be detected correctly even after applying attacks. Subsequently, the stego-system performance improved significantly against steganalysis attacks and data loss (Hamid et al., 2012a, 2013b; Abdul-mahdi et al., 2013).

A. Yahya, *Steganography Techniques for Digital Images*,
https://doi.org/10.1007/978-3-319-78597-4_3

3.2 Characteristic Region-Based Watermarking

In 2004, (Lowe, 2004) presented scale-invariant feature transform (SIFT) algorithm for extracting distinctive invariant features from images that can be invariant to image scale and rotation (Lowe, 2004). The algorithm was widely used in image mosaic, recognition, retrieval, and so on. Similar to SURF, SIFT not only detects interest points or topographies, but then again also proposes a technique to create an invariant descriptor. This descriptor can recognize the attentiveness points and contest them even under a diversity of disturbed conditions—for instance: scale changes, rotation, changes in illumination or viewpoints, or image noise (Bauer, Sünderhauf, & Protzel, 2006). A literature survey (Li, Qian, & Pan, 2011) found that SIFT is exploited for characteristic region detection to attain image watermark synchronization for copyright-protection reasons. Their system attained high-capacity data hiding and comprehensive watermark strength.

In this chapter, SIFT and SURF are independently used in a similar way to attain steganography synchronization (Hamid et al., 2012a). Then the two techniques are compared. The steganography synchronization algorithm consists of two stages, namely extraction of the robust interest points in the image and data hiding in the regions centered at these interest points. The regions are selected for concealing secret information to ensure that the positions of the areas in which the data are hidden can be recognized without an embedding map. Moreover, the areas in which the data are embedded are not fixed and are extremely reliant on the features of the image employed as a cover. Furthermore, selecting a few regions to hide data will diminish the alteration of the stego-image.

3.2.1 SIFT Detector

SIFT mainly includes four major stages, namely: scale-space extrema detection, interest point localization, orientation assignment, and interest point (feature point) descriptor. Given a digital image $I(x, y)$, its scale space representation, $L(x, y, \sigma)$ can be obtained by

$$L(x,y,\sigma) = I(x,y) * G(x,y,\sigma) \tag{3.1}$$

where $*$ is the convolution, and $G(x, y, \sigma)$ is the variable-scale Gaussian kernel with standard deviation σ.

The primary phase practices difference-of-Gaussian function (DoG) to recognize the possible attentiveness points, which are invariant to scale and orientation. DoG is used rather than Gaussian to increase the calculation speed (Juan & Gwun, 2009). The SIFT interest points are detected from the scale space of the image by finding the scale-space extrema in the DoG function, which can be obtained by subtracting two nearby scales separated by a constant multiplicative factor k (Lowe, 2004), as given by:

$$D(x,y,\sigma) = (G(x,y,k\sigma) - G(x,y,\sigma)) * I(x,y) = L(x,y,k\sigma) - L(x,y,\sigma) \quad (3.2)$$

In the interest point localization phase, the small distinction points are disallowed and the edge reply is eradicated. Hessian matrix was used to calculate the principal curvatures and remove the interest points that have ratios between the principal curvatures that are greater than a specific threshold. An orientation histogram was shaped from the gradient orientations of sample points inside an area near the interest point to obtain an orientation assignment (Juan & Gwun, 2009). According to their experiments, the best results were achieved with (4 × 4) arrays of histograms that each have 8 orientation bins. Hence, the SIFT used was (4 × 4 × 8 = 128) dimensions (Lowe, 2004). The interest point descriptors are planned from the local gradient orientation and magnitudes in a definite region near the known interest point. The gradient orientations and magnitudes are joint in a histogram illustration from which the descriptor is shaped (Bauer et al., 2006). The obtained descriptor is used for reliable image-matching. Accordingly, the SIFT detects an interest point with its coordinate (p_1, p_2), characteristic scale (σ), and orientation (θ). The characteristic scale is the scale in which the feature point is detected. Generally, the characteristic scale of a local structure is specified by a local extrema over scale of normalized derivatives (Li et al., 2011).

3.3 Theoretical Framework of the Proposed Scheme

In the information-hiding field, some researchers focus on methods to ensure robustness, such as watermarking, whereas others focus on methods to ensure imperceptibility, such as steganography. The current work presents a new steganographic system to achieve both robustness and imperceptibility. To achieve this goal, three techniques are exploited to develop the proposed algorithm, and each technique is used for a very specific function, as shown in Fig. 3.1. The three main components of the proposed CR-BIS are well-known algorithms and could be implemented in non-steganographic scenarios. However, the algorithms are combined, thereby exhibiting a distinctive interaction. The Blowfish encryption algorithm is adopted to encrypt the secret information prior to the embedding process (*Step 1*). SURF technique is used to identify the information-embedding regions in the image (*Step 2*). The Cohen-Daubechies-Feauveau (CDF) DWT is used to embed the encrypted payload in a content-based manner (*Step 3*). Figure 3.2 illustrates the block diagram of the proposed CR-BIS algorithm. The following sections of this chapter present a detailed explanation of the three main components of the CR-BIS algorithm and its development.

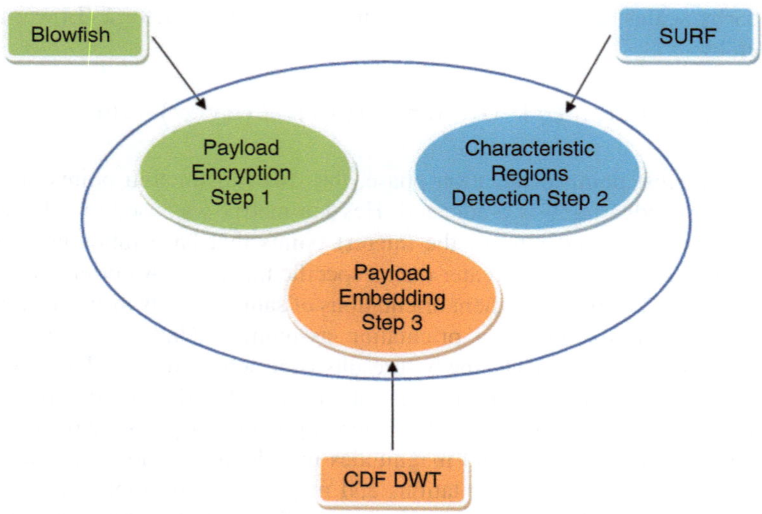

Fig. 3.1 The main components of the CR-BIS algorithm

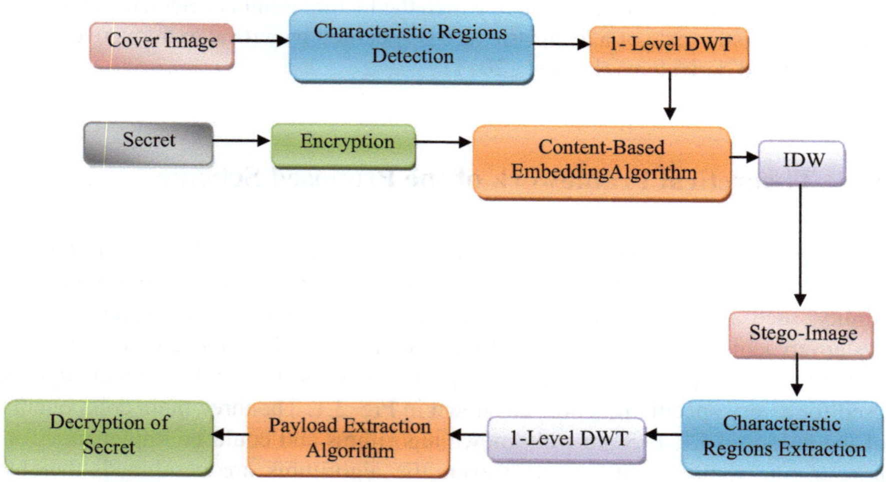

Fig. 3.2 Block diagram of the proposed CR-BIS method

3.3.1 *Payload Encryption*

A message is usually in plaintext. Encryption is the process of camouflaging a message to conceal its meaning. An encrypted message is referred to as ciphertext. Reverting ciphertext into its original form, such as plaintext, is known as decryption. Cryptography is the art and science of protecting messages by converting them into an unreadable format, and the conversion is accomplished by cryptographers. Cryptanalysts are specialists in cryptanalysis, which is the art and science of decoding ciphertext (Schneier, 2012).

A number of algorithms are used for message encryption. One such algorithm is the Blowfish algorithm, which was first designed by Bruce Schneier in 1993 to replace Data Encryption Standard (DES). The algorithm is a symmetric block cipher, and each block is 64 bits. The encryption algorithm is capable of managing keys from 8 bits to 448 bits in 8-bit steps (Schneier, 1994). This cryptographic algorithm is more efficient than other crypto algorithms and is suitable and well-organized for hardware implementation (Milad, Muda, Noh, & Algaet, 2012; Schneier, 1995; Tingyuan, Chuanwang, & Xulong, 2010; Verma & Singh, 2012). Blowfish does not involve cryptanalysis and is a part of the Linux kernel (Karthigai Kumar & Baskaran, 2010). The algorithm was designed to have the following features (Schneier, 1994):

Fast: It encrypts data on a 32-bit microprocessor at a rate of 18 clock cycles per byte.

Compact: It can run on less than 5 K of memory.

Simple: Blowfish's simple structure is easy to implement and simplifies the task of determining the efficiency of the algorithm.

Variable secure: The key length is variable and can be as long as 448 bits. This permits a tradeoff between higher speed and higher security. Blowfish encrypts 64-bit blocks of plaintext into 64-bit blocks of ciphertext. Moreover, it is frequently implemented in different fields and has received a reasonable level of security.

3.3.1.1 Blowfish Algorithm

The Blowfish algorithm is a block cipher, which means that it encrypts small pieces of data at a time before repeating the algorithm in the next data part of the entire file. This algorithm is performed by encrypting one 64-bit block chunk at a time, which is half of the block segment size in the Advanced Encryption Standard (AES) algorithm (Finch, 1995). The algorithm mainly involves two parts, namely key expansion and data encryption. The key-expansion stage converts a key with a maximum of 448 bits into a number of sub-key arrays totaling 4168 bytes. This key is indirectly used for encryption and is used to create many sub-keys with unknown patterns (Schneier, 1995).

Data encryption is implemented via a 16-round Feistel network. Each round consists of a key-dependent permutation and key- and data-dependent substitutions. A Feistel structure has many benefits, particularly in hardware, since to decrypt the ciphertext all that is needed is a reversal of the key schedule. All operations are exclusive-or-operation (XORs) and additions on 32-bit words. The only additional operations are four indexed array data lookups per round (Schneier, 1994).

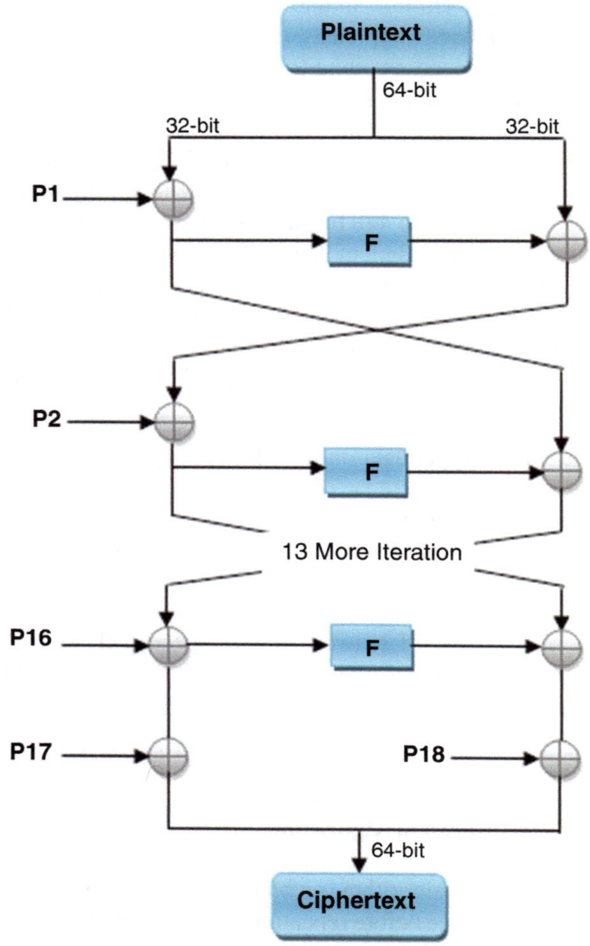

Fig. 3.3 The Blowfish algorithm. (Schneier, 1994)

3.3.1.2 Data Encryption

Data encryption starts with a 64-bit block element of plaintext, which will be converted into a 64-bit ciphertext. The 64-bit segment is divided into two equal halves to be used as the base of the Blowfish algorithm. The XOR is done between the first 32-bit block segment (XL) and the first P array, as shown in Fig. 3.3.

The resulting 32-bit data are passed on to the F function to permute the data and to produce a 32-bit block segment. This permuted block segment is XOR'ed with the second 32-bit segment (XR) created by the 64-bit plaintext splitting. After that,

XL and XR are swapped for future iterations of the Blowfish algorithm (Schneier, 1995). For further clarification, the 64-bit input plaintext is denoted by X, which is then divided into two equal halves, XL and XR. Then,

$$\text{For}\, i = 1 \text{ } to \text{ } 16$$

$$\text{XL} = \text{XL XOR P}i$$

$$\text{XR} = \text{F}\left(\text{XL}\right) \text{XOR XR}$$

$$\text{Swap XL and XR}$$

After the sixteenth iteration, XL and XR are swapped to cancel the last swap. Then XR = XR XOR P17 and XL = XL XOR P18. To obtain the required ciphertext, XL and XR are recombined (Schneier, 1994).

3.3.1.3 Key Schedule

The major strength of the Blowfish algorithm is its complex key schedule. Blowfish uses a large number of sub-keys that must be computed in advance before starting any data encryption or decryption (Schneier, 1995). The P-array includes eighteen 32-bit sub-keys, P1, P2, …., P18. Moreover, four 32-bit S-Boxes exist, each one with 256 entries:

$$S_{1,0};\ S_{1,1};\ldots\ldots S_{1,255}$$

$$\text{S}_{2,0};\ \text{S}_{2,1};\ldots\ldots\text{S}_{2,255}$$

$$S_{3,0};\ S_{3,1};\ldots\ldots S_{3,255}$$

$$S_{4,0};\ S_{4,1};\ldots\ldots S_{4,255}$$

The following steps describe sub-key generation (Schneier, 1995):

Step 1: First, the P-array and then the four S-Boxes are initialized in order with a fixed string. This string involves the hexadecimal digits of the fractional part of pi (π), which is currently supposed to have an undetectable pattern.

Step 2: XOR P1 with the first 32 bits of the key, XOR P2 with the second 32 bits of the key, and so on for all bits of the key (up to P14) are given. The cycle is repeated through the key bits until the entire P array has been XOR'ed with the key bits.

Step 3: Encrypt the all zero strings by using the Blowfish algorithm with the obtained keys from steps (1) and (2).

Step 4: P1 and P2 are exchanged with the output of step (3).

Fig. 3.4 The Feistel function of the Blowfish algorithm. (Schneier, 1994)

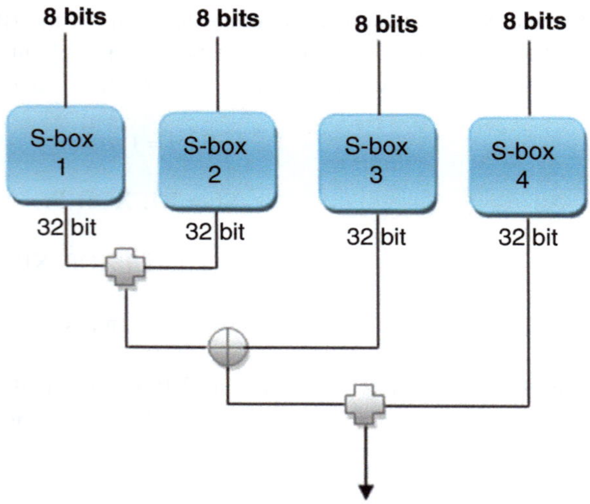

Step 5: The output of step (3) is encrypted by using the Blowfish algorithm with the modified sub-keys.

Step 6: P3 and P4 are replaced with the output of step (5).

Step 7: The procedures continue to be performed, replacing all entries of the P array, and then all four S-Boxes in order, with the output of the constantly altering Blowfish algorithm. To generate all the required sub-keys, 521 iterations are required because the P array is 576 bits long, and the key bytes are XOR'ed through all these 576 bits throughout the initialization. Many applications support key sizes of up to 576 bits. The key is restricted to 448 bits to confirm that every key of every sub-key depends on every bit of the key (Schneier, 1995).

3.3.1.4 F Function

The F function represents the most complex part of the Blowfish algorithm and is the only part that uses the S-Boxes. The contents of the function are given in Fig. 3.4. The F function splits the 32-bit stream of data into four 8-bit sections (a, b, c, and d). Each 8-bit quarter is transformed into a 32-bit data stream according to their corresponding S-Boxes. The S-Box outputs, which are 32-bit data streams, are added (modulo 2^{32}) and XOR'ed to produce the final 32-bit output, expressed by this equation:

$$\mathrm{F}(\mathrm{XL}) = \left(\left(S_{1,a} + S_{2,b} \bmod 2^{32}\right) \mathrm{XOR}\, S_{3,c}\right) + S_{4,d} \bmod 2^{32}$$

Decryption is the same as encryption, except that P1, P2, …, P18 are used in reverse order (Schneier, 1994).

3.3.1.5 Blowfish Algorithm Security and Performance

No effective cryptanalysis on the full-round version of Blowfish security has been published (Karthigai Kumar & Baskaran, 2010). In his PhD book, Rijmen introduced a second-order differential attack to break only four rounds of the Blowfish algorithm and no more (Daemen & Rijmen, 2002; Rijmen, 1997). No known method of breaking the full 16 rounds is available (Singh, Singla, & Sandha, 2012). The sub-key generation process, which is time-consuming, adds significant complexity for a brute-force attack. Very long sub-keys are difficult to store; thus, they should be generated by a brute-force cracking machine as required. The encryption algorithm requires a total of 522 iterations to test a single key, hence adding 29 steps to any brute-force attack (Schneier, 1994).

Singh and his colleagues (Singh et al., 2012) argued the performance of four algorithms, AES, DES, 3DES, and Blowfish. Their comparison is based on encryption time, decryption time, and throughput. Experimental results show that the Blowfish algorithm is more suitable for wireless networks. Moreover, their study reveals that Blowfish performs better than AES, DES, and 3DES in terms of encryption time, decryption time, and throughput. In 2012, Kumar and Karthikeyan showed that the Blowfish algorithm performs better than AES with different performance metrics. Their study shows that the Blowfish algorithm is good for text-based encryption, whereas AES is better for image encryption. For high-security and performance aspects, the Blowfish algorithm is highly recommended (M. A. Kumar & Karthikeyan, 2012).

3.3.2 *Identifying Embedding Regions Using SURF*

This section discusses the automatic identification of reliable regions in images to orient the embedding process. Most introduced steganographic techniques suffer from intolerance to geometric distortions applied to the stego-image. For example, if rotation or translation occurs, all the hidden data will be lost. This book introduces an object-oriented embedding approach to steganography to provide an automatic solution to different problems. Currently, the proposed scheme can be classified within adaptive steganography, which recognizes the textural or quasi-textural zones for embedding the secret information. The second attains the statistical universal structures of the image before making an effort to embed the secret data in specific sections of the image. These statistics will direct where to mark the fluctuations. In brief, information is embedded in special sections in the image depending on their characteristics. The same characteristics must be used to recognize the embedded sections appropriately to start the withdrawal procedure. This process requires that the characteristic credentials system must be strong enough to continue after potential threats, attacks, or communication errors. SURF is used because of its basic benefits, which contain invariance to rotation, translation, cropping, shifting, and fast automatic extraction of embedding regions.

Different techniques that are commonly used to extract image features or characteristics are SIFT (Lowe, 2004), SURF (Bay, 2006), Hu moments (Ming-Kuei, 1962), and Zernike moments (Teague, 1980). Each technique has its advantages and can be used for certain applications. In this book, the SURF technique is adopted to detect the interest points in the image. These interest points have the following properties (Bay, 2006):

- Accuracy of the interest point localization.
- Invariance to scale changes within a certain range.
- Invariance to rotations.
- Stability against changes in lighting and contrast.
- Steadiness toward affine or projective transformations.

SURF was published by (Bay, 2006) to tackle the problem of point and line segment correspondence between two images of a similar scene or piece. The second can be part of numerous computer vision applications. The SURF method can be distributed into three core stages.

First: Interest opinions are detected at typical positions in the image, for instance corners, blobs, and T-junctions. The most important property of the interest-point detector is its repeatability, which defines the reliability of a detector to find identical physical interest points under different viewing situations (Bay, 2006). Such points can be reliably extracted to provide a high amount of information. The interest-point detector is based on the Hessian matrix because of its good computation speed and accuracy. It detects blob-like structures in the locations where its determinant is maximum.

Given a point $\mathbf{x} = (x, y)$ in an image I, the Hessian matrix $\mathcal{H}(\boldsymbol{x}, \sigma)$ in $\mathbf{x}$ at scale σ is defined as follows:

$$\mathcal{H}(\boldsymbol{x},\sigma)=\begin{bmatrix} L_{xx}(\boldsymbol{x},\sigma) & L_{xy}(\boldsymbol{x},\sigma) \\ L_{xy}(\boldsymbol{x},\sigma) & L_{yy}(\boldsymbol{x},\sigma) \end{bmatrix} \tag{3.3}$$

where $L_{xx}(\boldsymbol{x}, \sigma)$ is the convolution of the Gaussian second-order derivative $\frac{\partial}{\partial x^2} g(\sigma)$ with the image I in point $\mathbf{x}$, and similarly for $L_{xy}(\boldsymbol{x}, \sigma)$ and $L_{yy}(\boldsymbol{x}, \sigma)$. Gaussians are considered the optimal choice for scale-space analysis (Koenderink, 1984; Lindeberg, 1990). The Hessian matrix is approximated by using simple box filters, such that the approximation for the second-order Gaussian partial derivative in the y-direction is denoted by D_{yy}. Similarly, the x-direction is indicated as D_{xx}, and the xy-direction is given as D_{xy}. Then, the determinant of the Hessian matrix will be given by the following equation:

$$\det\left(\mathcal{H}_{\text{approx}}\right) = D_{xx} D_{yy} - \left(0.9 D_{xy}\right)^2 \tag{3.4}$$

The approximated determinant of the Hessian matrix corresponds to the blob response in the image at location $\mathbf{x}$. These responses are stored in a blob response

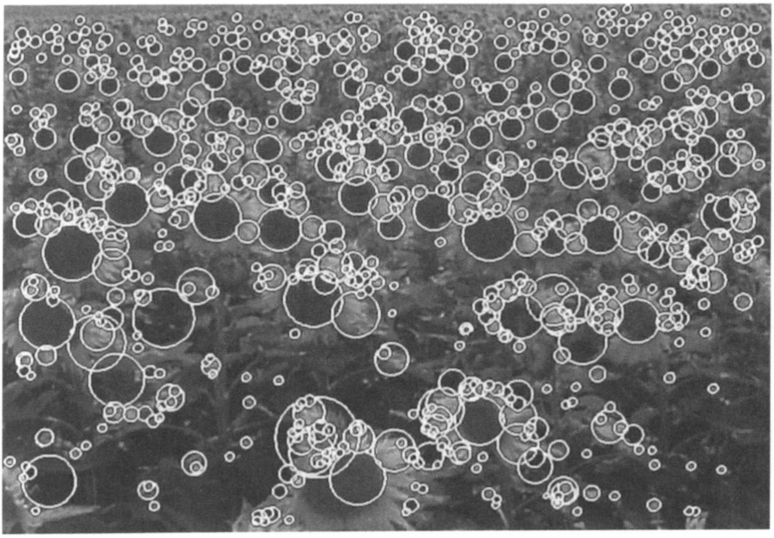

Fig. 3.5 Detected interest points for a sunflower field. (Bay, 2006)

map over different scales. Then the local maxima are detected to represent the detected interest points (Bay, 2006). Figure 3.5 shows an example of the detected interest points by using the described Fast-Hessian detector (Bay, 2006).

Second: The interest points obtained in the first step have to be robustly characterized by a descriptor to describe the distribution of the intensity content within the interest-point neighborhood, which is similar to the gradient information extracted by SIFT and its variants. This descriptor is based on the distribution of first-order Haar wavelet responses in x and y directions rather than the gradient. To be invariant to image rotation, a reproducible orientation for the interest points is identified. For that purpose, the Haar wavelet responses are calculated first in the x and y directions within a circular neighborhood of radius (6 s) around the interest point; (s) represents the scale in which the interest point is detected. To extract the descriptor, the first step includes the construction of a square region centered on the interest points and oriented along the orientation selected by the previous step. This square region is split into smaller 4 × 4 square sub-regions to retain significant spatial information. Then, for each sub-region, a few simple features are calculated at 5 × 5 regularly spaced sample points. After that, the wavelet responses in horizontal direction d_x and vertical direction d_y are summed over each sub-region to form the first set of entries to the feature vector. To obtain the information on the polarity of the intensity changes, the sum of the absolute values of the responses $|d_x|$ and $|d_y|$ is extracted. Accordingly, each sub-region has a four-dimensional descriptor vector $\boldsymbol{v}$ for its intensity structure. This extraction results in a descriptor vector for all 4 × 4 sub-regions of length 64 (Bay, 2006).

$$v = \sum\left(d_x + d_y + |d_x| + |d_y|\right) \quad (3.5)$$

Finally, the descriptor vectors are matched to different images. The matching is based on a distance between vectors, such as the Euclidean distance. In this stage, the features are only compared if they have the same type of contrast. This information facilitates faster matching without reducing the descriptor performance (Bay, 2006). This step is outside the scope of the present work.

3.3.3 Embedding Data Using CDF DWT

After generating the encrypted payload and identifying the robust invariant characteristic regions, the information-embedding phase is the important step to be accomplished. In addition to relying on the SURF algorithm to detect the robust regions in the image to restrict information, the present work proposes information embedding in the DWT domain to achieve high robustness against JPEG compression and noise. Furthermore, the information is embedded in a content-based manner (Li et al., 2011). DWT is a popular transformation method in the image-processing community, especially among experts in the area of image compression. Its applications in different areas are growing significantly. DWT adapts spatial domain data into the frequency domain data. Wavelets are employed in the image steganography model because the wavelet transform clearly partitions the high and low frequency information on a pixel. The DWT is preferred over the DCT due to its to the image at various levels (Kumar & Kumar, 2010). Wavelet has been chosen for the following reasons (Cheddad, 2009):

- The wavelet transform models the HVS better, more closely than DCT does.
- The visual effects that result in the wavelet-coded images are less obvious compared with that of DCT because the DWT does not decompose the image into blocks for processing.

Moreover, the DFT and the DCT are full-frame transforms such that any change in the transform coefficients affects the entire image unless that DCT is applied by using the block-based approach. Furthermore, DWT has spatial-frequency locality, which means that the embedded information will affect the image locally (Potdar, Han, & Chang, 2005). Consequently, a wavelet transform provides both frequency and spatial descriptions for the image. More verification can be found in (Silva & Agaian, 2004). The central focus of this chapter is the embedding of a secret message into the horizontal and vertical decomposition in the first- and second-level CDF DWT. The details of the proposed system for embedding secret information are explained in Sect. 3.4.1.

3.4 Developing the Proposed Algorithm

The proposed algorithm in this chapter is based on steganography synchronization (Hamid et al., 2012a,b, 2013a,b). Steganography synchronization confirms that information embedding and removal are fulfilled in a similar section. Further, most steganography systems require that the stego-image and its original version be obtainable to remove the data embedded in the stego-image; such schemes include statistical steganographic techniques, distortion techniques, several spatial domain methods, and some transform domain schemes. However, a reasonable steganography structure must enable its users to remove the hidden data from the stego-image without the assistance of its unique version (i.e., blind scheme) (Li, Li, & Wei, 2007). The research reported in this book introduces a completely blind stego-system because only the stego-image is required to initiate the extraction process.

Steganography synchronization is achieved via the characteristic regions that can be generated by using the SURF technique. SURF calculates a list of invariant interest points in the cover image. The detected interest facts will be the centers of the sections wherein the data is to be embedded. Depending on the size of the necessary sections, some points will not be used to evade any connections because of very close interest points. To guarantee disjoint local regions, each point should be measured by computing the Euclidian distance d. The calculation should be between the designated points and among all other points in the list. All d values must be greater than $2\sqrt{2}r$ because the size of the embedding section is given as ($2r \times 2r$) (see Fig. 3.6). In the same manner, if the required regions are circular in shape with a radius r, d should be greater than $2r$.

3.4.1 *Secret Data Embedding Phase*

Data embedding is implemented in the DWT domain for the reasons explained in Sect. 3.3.3. The comprehensive information embedding techniques are listed below.

1. The characteristic sections are removed from the cover image by means of SURF. At that time, the subsequent invariant interest points are examined to avoid any intersected regions; thus, some points are disregarded in this phase.
2. Using the last list of points, the embedded sections will be positioned in the cover image as circular sections with radius ($r = 64$) as shown in Fig. 3.7.
3. For individual embedded sections, first-level DWT on each characteristic section is functional to produce the wavelet coefficients. The 9/7 biorthogonal wavelet is adopted to implement the CR-BIS algorithm.
4. Horizontal and vertical high-frequency coefficients are scanned by using the raster method. Then the information bits are embedded by adjusting the horizontal and vertical coefficients in a content-based manner, as explained in Fig. 3.8.
5. For a payload bit b,the corresponding horizontal and vertical wavelet coefficients are first selected and denoted by $H(x,y)$ and $V(x,y)$, respectively. Then b is

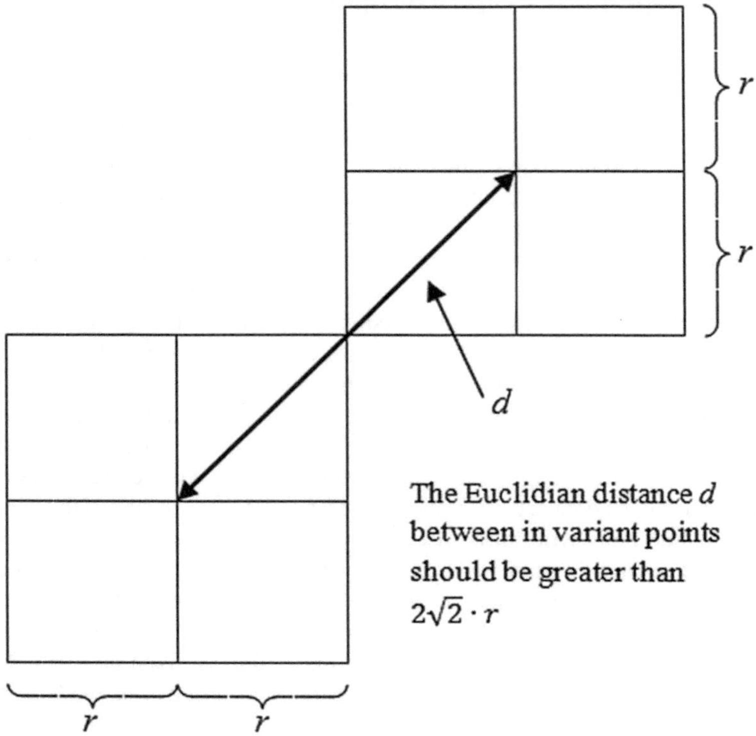

Fig. 3.6 Examining characteristic regions to avoid intersections

Fig. 3.7 Detecting characteristic regions using the SURF technique

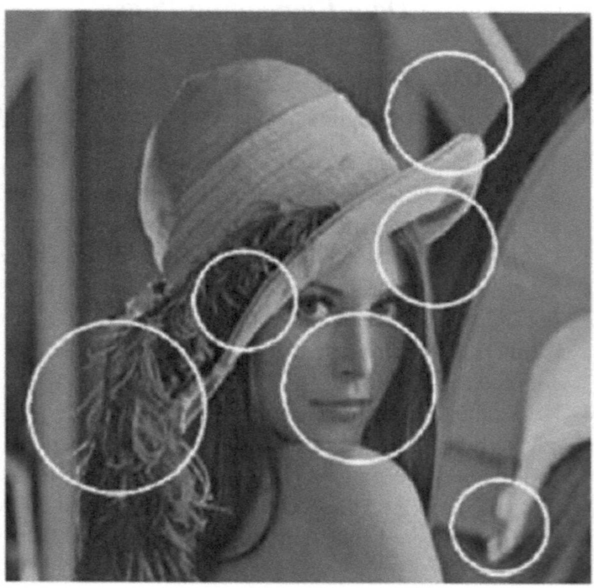

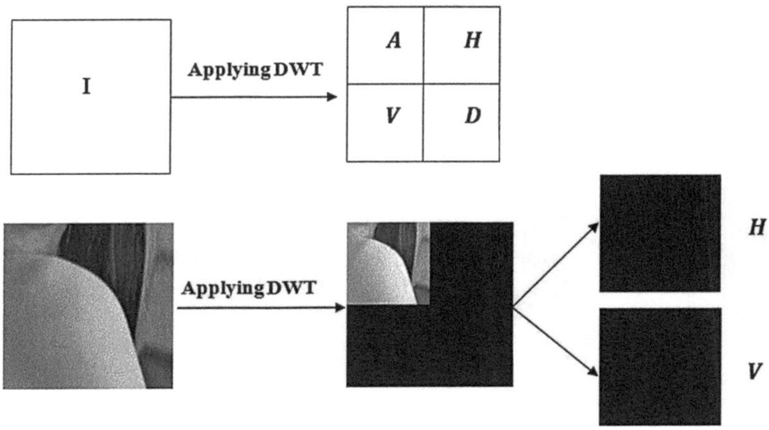

Fig. 3.8 Decomposing the image into four sub-bands using DWT

embedded by increasing the difference between $H(x, y)$ and $V(x, y)$. The rules of wavelet coefficient modification are as follows:

6. If $b = 1$ and $D_1 = H(x, y) - V(x, y) < T$ (T is a threshold to control the information invisibility), $H(x, y)$ should be increased while decreasing $V(x, y)$. The stage is completed by injecting the secret message as specified below:

$$\begin{cases} H^{\sim}(x,y) = H(x,y) + \dfrac{T - D_1}{2} \\ V^{\sim}(x,y) = V(x,y) - \dfrac{T - D_1}{2} \end{cases} \tag{3.6}$$

Otherwise, if $D_1 = H(x, y) - V(x, y) \geq T$, no further step is performed. If $b = 0$ and $D_2 = V(x, y) - H(x, y) < T$, the same process is implemented.

$$\begin{cases} H^{\sim}(x,y) = H(x,y) - \dfrac{T - D_2}{2} \\ V^{\sim}(x,y) = V(x,y) + \dfrac{T - D_2}{2} \end{cases} \tag{3.7}$$

Otherwise, if $D_2 = V(x, y) - H(x, y) \geq T$, no further step is performed. Lastly, one-level inverse DWT is practical to attain the stego square section. Later, the original characteristic section is substituted with the stego one. The whole embedding stage is demonstrated in Fig. 3.8. For the complete removed characteristic sections, the aforesaid embedding processes are repeated to produce the entire stego-image. The embedding phase is illustrated in Fig. 3.9. Figure 3.10 showcases the flowchart of the CR-BIS algorithm.

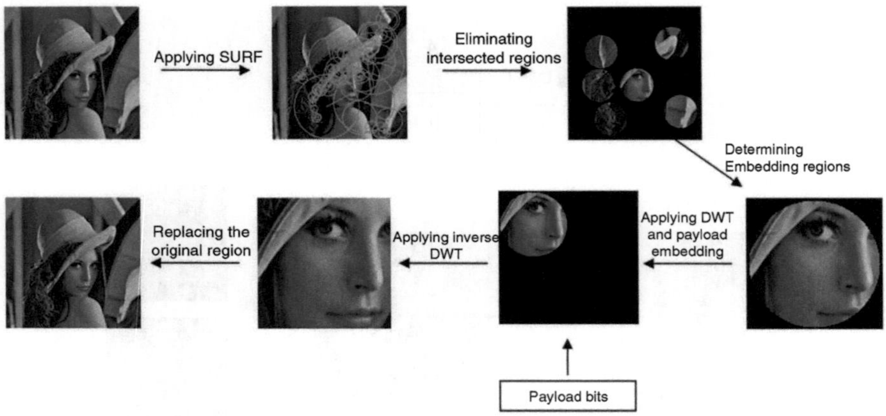

Fig. 3.9 The embedding phase of the CR-BIS algorithm

3.4.2 Secret Data Extracting Stage

The first two stages of the information-removal stage are precisely the same as information embedding. Characteristic sections are first removed from the probably distorted image, using the SURF system (Hamid et al., 2013b). The invariant interest points are inspected to evade interconnected sections. Then the embedded sections are firm. Afterward, the payload removal is completed on each local section as demonstrated below:

1. One level DWT is applied on each invariant region to obtain the wavelet coefficients.
2. The horizontal and vertical coefficients are determined and represented by $H(x, y)$ and $V(x, y)$, respectively. Then each payload bit b can be removed by relating the resultant horizontal and vertical coefficients, as shown in Eq. 3.8.

$$b = \begin{cases} 1 \, if \, H(x,y) > V(x,y) \\ 0 \, if \, H(x,y) < V(x,y) \end{cases} \tag{3.8}$$

3.5 Computer Simulation and Results

To assess the performance of the proposed scheme, experiments were conducted on 10 standard 512 × 512-pixel grayscale images, as shown in Fig. 3.11 (Hamid et al., 2013b). The payload is a randomly generated binary data. For comparison, first-level and second-level 9/7 biorthogonal wavelets were used. The threshold (T) employed for payload embedding is set to 1, which is determined by tests.

(a)

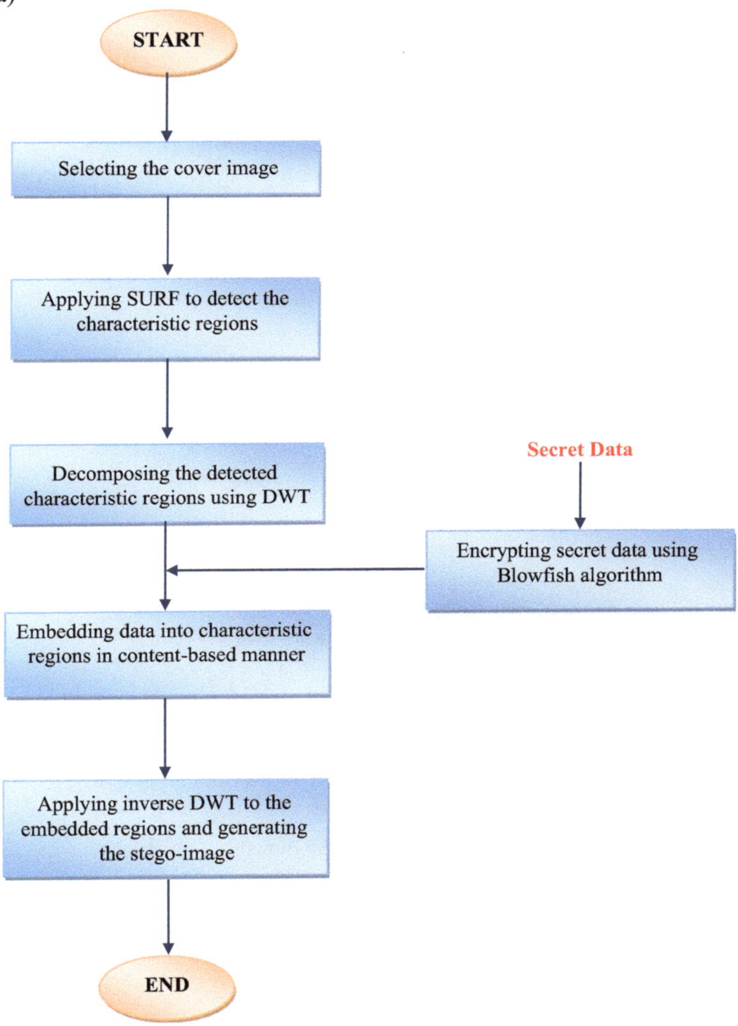

Fig. 3.10 Flowchart of the proposed CR-BIS algorithm (**a**) Secret data embedding phase (**b**) Secret data extraction phase

(b)

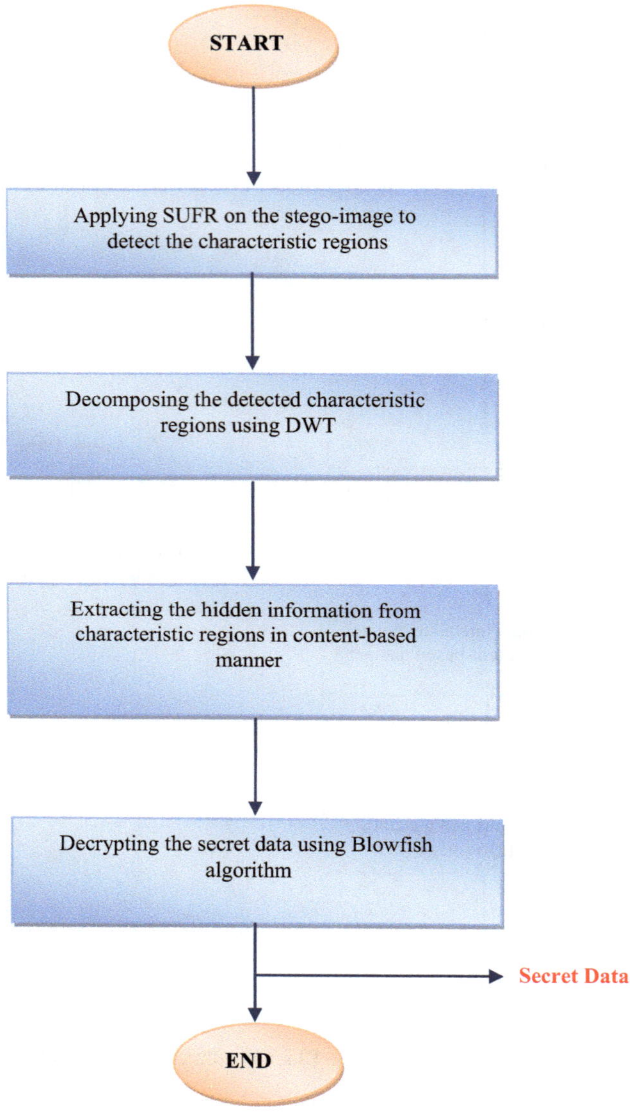

Fig. 3.10 (continued)

To check the strength of the recommended system, diverse attacks of changed stages are implemented on the stego-images. The attacks involved are JPEG compression with different quality factor (QF) values, Additive White Gaussian Noise (AWGN), and salt and pepper noise, as shown in Fig. 3.12 for the image "Lena." For evaluation purposes, different values of the above-mentioned attacks

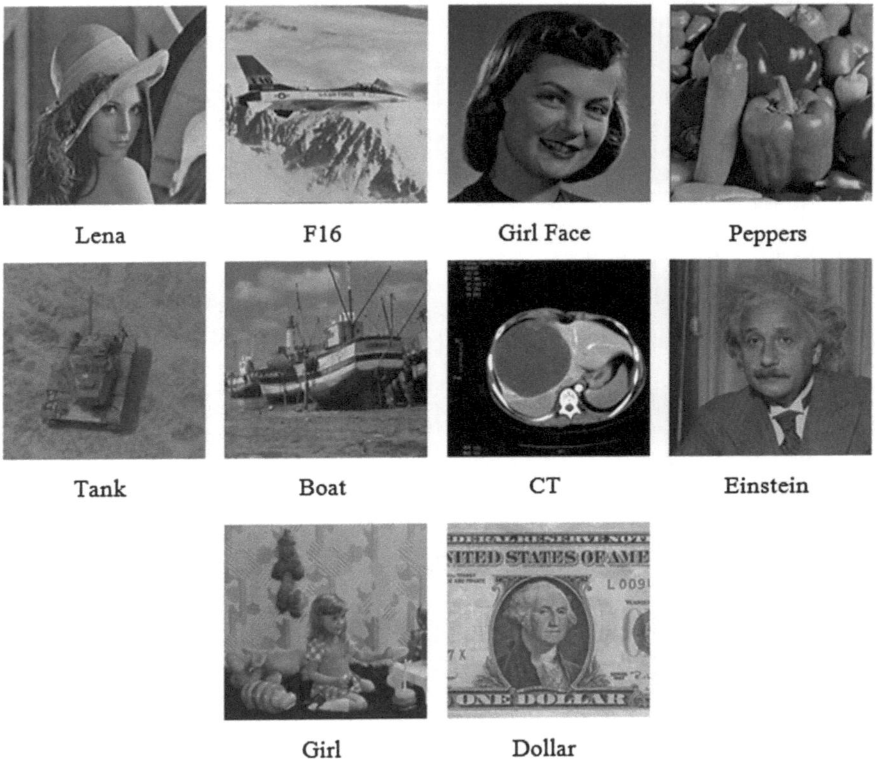

Fig. 3.11 Standard images used for evaluation

are applied to the stego-image. Then the extracted payload is compared with the embedded payload, and the BER is calculated as follows:

$$\text{BER} = \frac{\text{Number of error bits}}{\text{Total number of embedded bits}} \tag{3.9}$$

Besides the BER, the precision of synchronization (precision of properly perceived characteristic sections, denoted by ADR) is measured by means of SURF. To accomplish this stage, the investigators compute the percentage of the number of sections that were correctly recognized throughout the removal stage. For each kind of attack, the procedure is repeated by using 100 data files, and the averages are calculated. For comparison, the CR-BIS algorithm is implemented by using first- and second-level DWT separately.

Moreover, another synchronization-based Steganographic algorithm, which was proposed by (Li et al., 2011), is implemented in the present work for comparison purposes. Li et al.'s algorithm exploits SIFT for characteristic region detection.

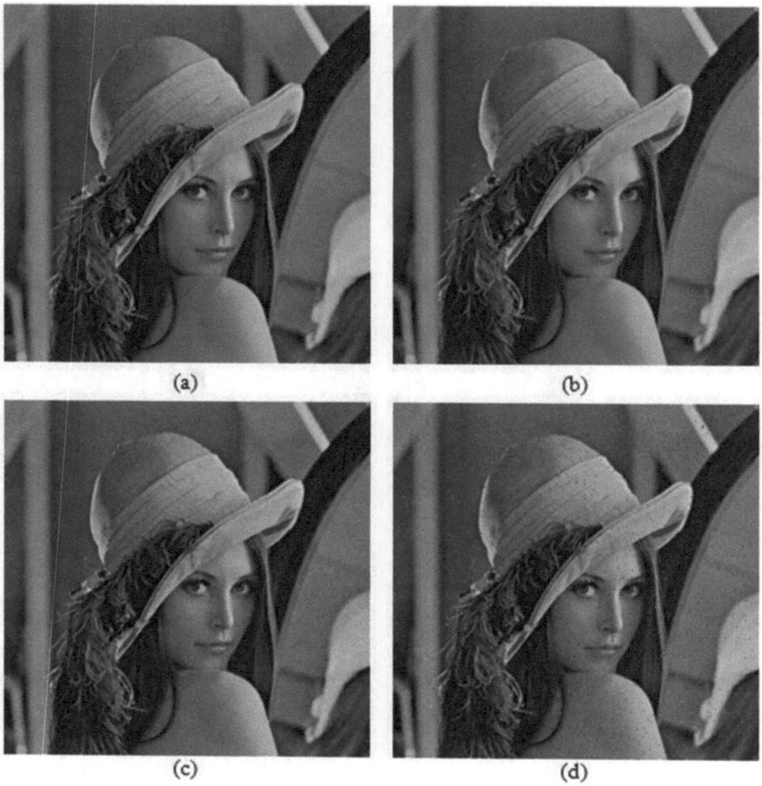

Fig. 3.12 The stego-image (**a**) Before attack. (**b**) After compression (QF = 70%) (**c**) After addition of Gaussian noise (SNR = 25 dB) (**d**) After addition of salt and pepper noise (SNR = 20 dB)

Furthermore, first- and second-level DWTs are used separately for embedding, and the same 100 data files are used during embedding. A comparison between the proposed algorithm and Li et al.'s algorithm is illustrated in Tables 3.1, 3.2, 3.3, 3.4, 3.5, 3.6, 3.7, 3.8, 3.9, and 3.10 for each individual image. Figures 3.13, 3.14, 3.15, 3.16, 3.17, 3.18. 3.19, 3.20, 3.21, 3.22, 3.23, and 3.24 compare the performances of the proposed algorithm and the (Li et al., 2011) algorithm in terms of ADR and BER for each type of attack. Table 3.11 compares the proposed CR-BIS algorithm and Li's algorithm in terms of the average of ADR and BER for all images used for testing. The comparison is also illustrated in Fig. 3.25.

In addition to robustness and reliability, which are measured by ADR and BER, the proposed scheme is evaluated in terms of the hiding capability and visual value of the stego-image. The capability is calculated by the number of payload bits that can be embedded in the image. Alternatively, the visual quality is calculated through the (PSNR), as given in Eq. (2.6) of (Cover, 2006).

The hiding capacity and the corresponding PSNR values achieved by using the proposed algorithm are shown in Table 3.12. However, a comparison between the hiding capacities of the two algorithms is inadequate, because the capacities of

Table 3.1 A comparison between Li's algorithm and the CR-BIS algorithm in terms of ADR and BER for the image "Lena"

	Li' algorithm				The proposed algorithm			
	1-level DWT		2-level DWT		1-level DWT		2-level DWT	
Type of attack	ADR (%)	BER (%)	ADR (%)	BER (%)	ADR (%)	BER (%)	ADR (%)	BER (%)
No attack	87.4	6.40	83.80	5.52	100	0	97.33	0.94
JPEG compression (QF = 100%)	84.2	8.02	78.00	7.54	100	0.07	97.00	1.05
JPEG compression (QF = 90%)	41.4	39.67	42.20	23.44	100	31.03	96.50	3.70
JPEG compression (QF = 80%)	37.6	44.74	45.80	26.99	99.83	39.84	95.67	13.66
AWGN (45 dB)	60.6	20.10	58.20	14.78	100	2.86	96.33	1.37
AWGN (35 dB)	32.8	36.21	35.60	27.73	98.17	21.95	92.83	10.55
AWGN (25 dB)	12.2	43.36	14.40	40.80	73.67	37.97	72.83	30.25
Salt & pepper noise (30 dB)	77.00	10.79	74.00	9.56	99.17	0.21	96.67	1.24
Salt & pepper noise (25 dB)	18.00	34.88	19.20	34.19	75.33	13.02	73.67	12.96
Salt & pepper noise (20 dB)	8.40	40.42	10.20	40.12	52.83	24.68	51.17	24.98
Median filter (3 × 3)	34.2	47.29	40.60	32.66	83.33	47.02	83.33	23.21
Low pass filter (3 × 3)	48.4	48.36	53.00	25.99	100	49.17	99.33	18.88

Table 3.2 A comparison between Li's algorithm and the CR-BIS algorithm in terms of ADR and BER for image "F16"

	Li' algorithm				The proposed algorithm			
	1-level DWT		2-level DWT		1-level DWT		2-level DWT	
Type of attack	ADR (%)	BER (%)	ADR (%)	BER (%)	ADR (%)	BER (%)	ADR (%)	BER (%)
No attack	83.50	5.20	82.17	6.18	100	0	99.5	0.37
JPEG compression (QF = 100%)	83.33	6.40	80.33	6.16	100	0.06	99.5	0.37
JPEG compression (QF = 90%)	76.50	31.84	72.33	10.32	100	31.75	99.5	2.62
JPEG compression (QF = 80%)	61.83	38.13	56.33	22.78	100	40.59	99.5	12.82
AWGN (45 dB)	71.83	15.00	73.83	8.89	100	2.93	99.5	0.40
AWGN (35 dB)	34.01	38.96	53.83	20.36	99.50	22.04	99	8.87
AWGN (25 dB)	26.17	49.67	36.17	32.47	78.75	40.11	80.25	30.30
Salt & pepper noise (30 dB)	50.17	19.06	44.67	20.39	87.25	7.91	88.5	7.12
Salt & pepper noise (25 dB)	30.50	28.17	29.67	27.04	66.00	20.86	65.75	20.22
Salt & pepper noise (20 dB)	18.33	33.08	22.50	30.48	48.25	33.86	49.25	33.86
Median filter (3 × 3)	29.67	44.56	25.50	35.48	49.50	48.25	46.50	30.65
Low pass filter (3 × 3)	50.67	76.33	55.17	31.88	100	49.60	99.75	24.13

Table 3.3 A comparison between Li's algorithm and the CR-BIS algorithm in terms of ADR and BER for the image "Girl Face"

	Li' algorithm				The proposed algorithm			
	1-level DWT		2-level DWT		1-level DWT		2-level DWT	
Type of attack	ADR (%)	BER (%)	ADR (%)	BER (%)	ADR (%)	BER (%)	ADR (%)	BER (%)
No attack	99.71	1.27	84.43	5.56	100	0.90	99.83	2.50
JPEG compression (QF = 100%)	98.71	1.65	84.00	5.57	100	1.05	99.67	2.59
JPEG compression (QF = 90%)	92.29	31.75	77.00	10.76	100	32.25	99.67	5.62
JPEG compression (QF = 80%)	63.86	40.31	67.00	19.42	99.83	40.09	99.67	14.77
AWGN (45 dB)	89.57	5.44	77.86	7.28	100	3.49	99.83	2.77
AGWN (35 dB)	63.71	21.38	57.14	17.99	99.33	21.14	98.83	10.13
AWGN (25 dB)	26.14	37.78	19.86	35.77	82.83	39.29	81.33	29.91
Salt & pepper noise (30 dB)	78.57	7.47	63.71	12.11	98.67	2.12	96.67	4.51
Salt & pepper noise (25 dB)	52.00	17.97	41.00	20.74	90.00	7.20	88.67	9.30
Salt & pepper noise (20 dB)	26.86	29.75	26.14	29.47	69.50	18.44	71.50	18.77
Median filter (3 × 3)	37.71	34.31	38.43	23.17	93.5	45.82	84.17	23.30
Low pass filter (3 × 3)	28.57	35.41	28.14	25.27	100	46.83	99.83	18.83

Table 3.4 A comparison between Li's algorithm and the CR-BIS algorithm in terms of ADR and BER for the image "Peppers"

	Li' algorithm				The proposed algorithm			
	1-level DWT		2-level DWT		1-level DWT		2-level DWT	
Type of attack	ADR (%)	BER (%)	ADR (%)	BER (%)	ADR (%)	BER (%)	ADR (%)	BER (%)
No attack	91.5	4.74	90.75	4.48	100	0.28	98.25	1.35
JPEG compression (QF = 100%)	89	6.05	85.50	6.70	100	0.36	98.00	1.42
JPEG compression (QF = 90%)	76.25	34.87	72.50	15.49	100	31.44	98.25	3.90
JPEG compression (QF = 80%)	75.75	41.66	69.25	22.66	91.25	40.57	84.00	16.67
AWGN (45 dB)	77.25	13.90	73.75	12.60	100	3.02	97.00	1.80
AWGN (35 dB)	66	30.88	65.25	22.33	97.00	22.14	94.50	10.87
AWGN (25 dB)	45.75	44.08	43.25	40.52	90.25	37.34	88.00	31.55
Salt & pepper noise (30 dB)	63	19.29	63.75	18.19	96.75	2.77	92.25	4.89
Salt & pepper noise (25 dB)	51.25	25.83	45.5	28.22	88	8.82	86.25	10.01
Salt & pepper noise (20 dB)	34.75	35.44	35	36.43	83	15.62	82.25	18.91
Median filter (3 × 3)	69.75	49.01	50.00	34.54	75.75	49.43	75.00	22.92
Low pass filter (3 × 3)	52.5	51.49	82.25	27.85	100	53.32	100	16.51

Table 3.5 A comparison between Li's algorithm and the CR-BIS algorithm in terms of ADR and BER for the image "Tank"

	Li' algorithm				The proposed algorithm			
	1-level DWT		2-level DWT		1-level DWT		2-level DWT	
Type of attack	ADR (%)	BER (%)	ADR (%)	BER (%)	ADR (%)	BER (%)	ADR (%)	BER (%)
No attack	85.00	4.43	76.33	6.20	100	0	98.5	0.56
JPEG compression (QF = 100%)	84.67	4.81	80.33	5.43	100	0.06	98.25	0.64
JPEG compression (QF = 90%)	74.33	29.17	67.50	11.27	100	28.72	98.75	2.97
JPEG compression (QF = 80%)	62.17	37.65	58.83	25.01	100	38.24	98.25	12.17
AWGN (45 dB)	75.50	9.48	68.00	9.60	100	1.10	98.25	0.65
AWGN (35 dB)	51.83	30.73	61.00	15.59	100	15.26	97.25	5.12
AWGN (25 dB)	15.33	39.84	30.83	32.06	93.25	33.59	83.5	24.93
Salt & pepper noise (30 dB)	49.17	23.80	46.00	23.94	96	3.29	90.5	5.71
Salt & pepper noise (25 dB)	49.67	24.04	46.67	22.67	97	2.80	90.75	5.69
Salt & pepper noise (20 dB)	9.83	39.51	9.17	39.26	66.75	24.32	60.25	26.96
Median filter (3 × 3)	26.50	46.31	34.00	40.37	69.50	48.83	58	30.37
Low pass filter (3 × 3)	97.00	51.27	93.83	23.35	100	48.77	96.75	25.64

Table 3.6 A comparison between Li's algorithm and the CR-BIS algorithm in terms of ADR and BER for the image "Boat"

	Li' algorithm				The proposed algorithm			
	1-level DWT		2-level DWT		1-level DWT		2-level DWT	
Type of attack	ADR (%)	BER (%)	ADR (%)	BER (%)	ADR (%)	BER (%)	ADR (%)	BER (%)
No attack	100	0.24	99.6	0.75	100	0.11	76	9.43
JPEG compression (QF = 100%)	100	0.30	99.4	0.83	100	0.16	76.25	9.32
JPEG compression (QF = 90%)	79.2	33.70	78.8	12.91	99.75	29.65	76	11.47
JPEG compression (QF = 80%)	91.4	39.06	91.6	15.68	99.25	38.86	75.5	19.59
AWGN (45 dB)	98.40	3.55	96.2	2.38	100	1.94	76.5	9.24
AWGN (35 dB)	87.60	19.08	83.8	12.38	99	19.00	76.25	15.14
AWGN (25 dB)	48.00	38.92	48.8	33.63	69.5	39.32	58.5	34.58
Salt & pepper noise (30 dB)	93.20	2.96	93.60	3.33	99.75	0.37	76.5	9.44
Salt & pepper noise (25 dB)	45.40	26.01	49.00	24.18	64.25	20.14	55.50	23.16
Salt & pepper noise (20 dB)	20.00	38.58	24.00	35.55	45.75	31.59	41.50	33.06
Median filter (3 × 3)	60.60	49.21	60.8	37.55	95.75	49.39	75.00	29.10
Low pass filter (3 × 3)	100	48.82	99.8	30.02	75	50.44	74.50	26.83

Table 3.7 A comparison between Li's algorithm and the CR-BIS algorithm in terms of ADR and BER for the image "CT"

	Li' algorithm				The proposed algorithm			
	1-level DWT		2-level DWT		1-level DWT		2-level DWT	
Type of attack	ADR (%)	BER (%)	ADR (%)	BER (%)	ADR (%)	BER (%)	ADR (%)	BER (%)
No attack	99.6	5.19	73.2	16.11	100	5.11	100	9.84
JPEG compression (QF = 100%)	99	6.03	72.4	16.52	100	5.66	100	9.96
JPEG compression (QF = 90%)	94.8	36.25	68.6	22.91	100	35.73	100	17.11
JPEG compression (QF = 80%)	96	42.44	57.4	31.08	100	42.10	100	26.87
AWGN (45 dB)	96	10.56	73.2	16.51	100	8.32	100	10.61
AWGN (35 dB)	93	22.37	70.4	22.44	100	21.82	100	16.73
AWGN (25 dB)	67.8	35.71	42.6	33.62	99.33	34.81	99.33	29.23
Salt & pepper noise (30 dB)	93.8	7.77	68.6	17.64	100	5.36	100	10.29
Salt & pepper noise (25 dB)	81.4	13.69	65	19.43	99.17	6.56	99.17	11.77
Salt & pepper noise (20 dB)	57.4	24.54	48.6	26.26	93.67	11.43	93.00	16.70
Median filter (3 × 3)	52.4	48.30	46.8	31.36	99.67	49.52	99.00	26.07
Low pass filter (3 × 3)	95.8	52.16	55	30.73	100	52.46	100	25.98

Table 3.8 A comparison between Li's algorithm and the CR-BIS algorithm in terms of ADR and BER for the image "Einstein"

	Li' algorithm				The proposed algorithm			
	1-level DWT		2-level DWT		1-level DWT		2-level DWT	
Type of attack	ADR (%)	BER (%)	ADR (%)	BER (%)	ADR (%)	BER (%)	ADR (%)	BER (%)
No attack	99.00	0.49	98.20	0.81	98.40	0.80	87.80	3.50
JPEG compression (QF = 100%)	96.60	1.74	97.40	1.25	97.80	1.15	87.20	3.71
JPEG compression (QF = 90%)	90.40	39.86	89.80	8.19	96.20	38.83	87.00	7.85
JPEG compression (QF = 80%)	93.60	47.34	86.00	8.34	97.60	46.95	86.20	22.42
AWGN (45 dB)	87.80	6.35	87.40	4.71	97.20	3.02	86.80	3.55
AWGN (35 dB)	50.00	29.23	47.00	21.53	95.00	19.76	87.20	9.19
AWGN (25 dB)	26.00	35.73	25.80	41.54	79.20	37.16	74.80	28.07
Salt & pepper noise (30 dB)	58.40	15.54	59.80	15.15	91.80	4.54	85.20	5.58
Salt & pepper noise (25 dB)	36.80	28.00	30.80	28.13	79.80	10.79	72.40	13.29
Salt & pepper noise (20 dB)	20.60	37.24	20.80	35.71	56.20	24.85	58.60	24.73
Median filter (3 × 3)	89.40	47.06	67.20	27.63	94.60	24.31	94.60	24.31
Low pass filter (3 × 3)	83.60	49.86	81.80	26.14	96.40	49.50	84.80	20.32

Table 3.9 A comparison between Li's algorithm and the CR-BIS algorithm in terms of ADR and BER for the image "Girl"

	Li' algorithm				The proposed algorithm			
	1-level DWT		2-level DWT		1-level DWT		2-level DWT	
Type of attack	ADR (%)	BER (%)	ADR (%)	BER (%)	ADR (%)	BER (%)	ADR (%)	BER (%)
No attack	99	0.50	95.43	0.55	100	0	97.25	0.95
JPEG compression (QF = 100%)	97.57	1.27	94.71	0.77	100	0.06	96.50	1.22
JPEG compression (QF = 90%)	85.71	35.62	85.14	6.80	100	32.09	96.00	3.73
JPEG compression (QF = 80%)	71.43	43.82	79.14	17.74	91.50	40.92	92.75	14.02
AWGN (45 dB)	89.29	8.66	86.71	4.12	100	3.52	96.75	1.18
AWGN (35 dB)	62.71	32.40	59.86	21.38	95.75	24.09	90.25	12.42
AWGN (25 dB)	19.86	39.47	20.14	36.56	88.75	39.52	87.00	29.23
Salt & pepper noise (30 dB)	58.43	18.47	54.43	16.85	95.25	3.54	94.00	4.15
Salt & pepper noise (25 dB)	50.71	22.73	50.43	19.67	90.75	5.99	90.25	5.87
Salt & pepper noise (20 dB)	9.43	37.36	8.71	34.93	78.75	19.33	77.25	22.34
Median filter (3 × 3)	75.57	46.73	74.14	23.31	100	24.72	100	24.72
Low pass filter (3 × 3)	31.71	49.53	30.29	35.70	81.25	49.59	82.75	25.21

Table 3.10 A comparison between Li's algorithm and the CR-BIS algorithm in terms of ADR and BER for the image "Dollar"

	Li' algorithm				The proposed algorithm			
	1-level DWT		2-level DWT		1-level DWT		2-level DWT	
Type of attack	ADR (%)	BER (%)	ADR (%)	BER (%)	ADR (%)	BER (%)	ADR (%)	BER (%)
No attack	99.75	0.47	94.00	3.13	98.83	0.70	76.17	10.17
JPEG compression (QF = 100%)	99.75	0.52	94.00	3.14	98.67	0.83	76.33	10.11
JPEG compression (QF = 90%)	96.50	21.73	85.00	24.03	95.67	23.78	75.83	12.19
JPEG compression (QF = 80%)	91.25	29.00	84.00	16.24	93.17	30.87	76.33	18.31
AWGN (45 dB)	96.25	6.16	87.50	5.82	98.83	5.10	76.17	10.26
AWGN (35 dB)	85.25	23.15	80.00	16.61	89.00	23.16	71.50	19.79
AWGN (25 dB)	53.25	36.92	53.00	31.71	64.33	36.78	56.83	31.46
Salt & pepper noise (30 dB)	78.50	11.38	76.00	11.88	76.33	12.99	61.50	17.16
Salt & pepper noise (25 dB)	40.75	31.93	40.50	31.39	80.33	10.92	64.00	15.99
Salt & pepper noise (20 dB)	36.75	33.32	36.25	32.45	49.33	29.14	42.17	30.35
Median filter (3 × 3)	47.00	51.08	64.00	40.06	24.83	40.59	24.83	40.59
Low pass filter (3 × 3)	83.75	47.20	55.50	35.19	57.17	47.27	52.33	34.87

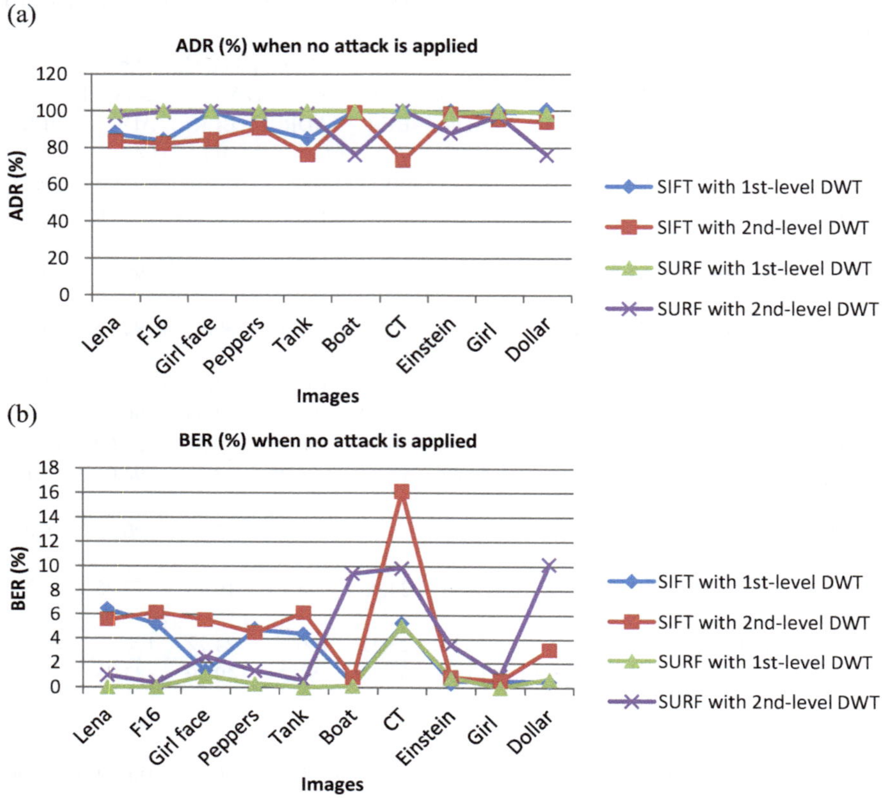

Fig. 3.13 Performance comparison in terms of (**a**) ADR and (**b**) BER when no attack is applied

the two algorithms depend on the number of detected characteristic regions in an image. The numbers of the characteristic regions which can be detected by using SIFT or SURF relies on user-defined thresholds. For certain values of thresholds, SIFT may produce a higher hiding capacity, while for another set of thresholds SURF may produce a higher capacity. The same can be said in the case of PSNR, which is inversely proportional to the hiding capacity.

3.6 Discussion

In this chapter, a novel image steganography scheme is proposed to achieve high robustness and a reliable information-hiding scheme and invariance to certain signal processing attacks, such as JPEG compression, AWGN, and salt and pepper noise. In this study, steganography synchronization is accomplished by means of the local

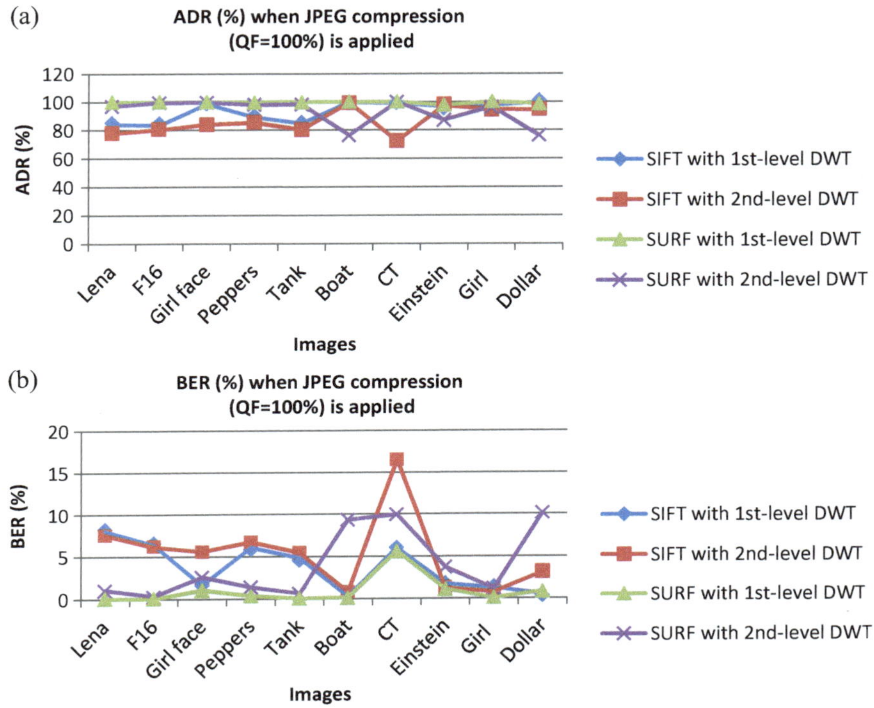

Fig. 3.14 Performance comparison in terms of (**a**) ADR and (**b**) BER when JPEG compression (QF = 100%) is applied

characteristic sections, whereas the secret binary data is embedded into the characteristic sections in the DWT domain in a content-based way. Employing characteristic regions provides a dynamic image-dependent manner of embedding data, compared with algorithms which embed data into predefined regions (Cheddad et al., 2010; Cheddad, Condell, Curran, & Mc Kevitt, 2009; Cheddad, Condell, Curran, & McKevitt, 2008a,b; Li et al., 2011). Such techniques add more security, as the locations in which the data are hidden depend on the image used and on its details (features).

The experimental outcomes display that the offered system can fight the signal processing threats used for evaluation. When first-level DWT is used for embedding, the SURF technique can detect the feature points with an average precision of up to 99.72% and 100% in all images except "Einstein," as shown in Tables 3.1, 3.2, 3.3, 3.4, 3.5, 3.6, 3.7, 3.8, 3.9, 3.10, and 3.11. However, the accuracy decreases dramatically when high Gaussian noise is applied, as shown in Figs. 3.17, 3.18, and 3.19, respectively. Gaussian noise affects the whole image because of its nature and can be mathematically tractable in both spatial and frequency domains. Such noise changes the image pixel values globally, and SURF is unable to detect the same

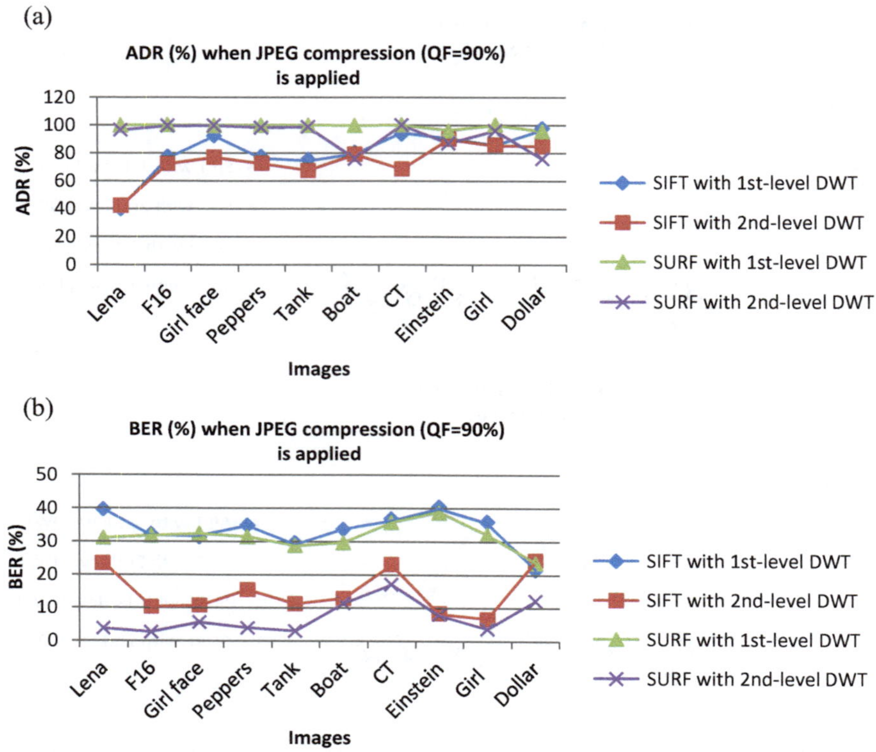

Fig. 3.15 Performance comparison in terms of (**a**) ADR and (**b**) BER when JPEG compression (QF = 90%) is applied

robust regions which were detected previously during the embedding phase. As a result, the high level of Gaussian noise decreases the ADR values.

On the other hand, exploiting second-level DWT for embedding enhances BER against attacks, but affects the quality of the stego-image as the PSNR values are reduced, as shown in Table 3.12. Likewise, second-level DWT reduces the accuracy of SURF as shown in Tables 3.1, 3.2, 3.3, 3.4, 3.5, 3.6, 3.7, 3.8, 3.9, 3.10, and 3.11 and illustrated in Figs. 3.13, 3.14, 3.15, 3.16, 3.17, 3.18. 3.19, 3.20, 3.21, 3.22, 3.23, and 3.24. Utilizing higher DWT levels for embedding changes the pixel values of the stego-image significantly. Consequently, SURF is unable to detect the same robust regions which were detected during the embedding phase and consequently decreases the ADR values. However, the BER values upgraded expressively when second-level DWT is used, except for the salt and pepper noise, which relatively increased the BER values since the correctness of the characteristic region detection is affected negatively.

The hiding capability attained by the offered structure is 0.04–0.06 bit per pixel (bpp) and (0.009–0.012 bpp) by using first-level DWT and second-level DWT,

(a)

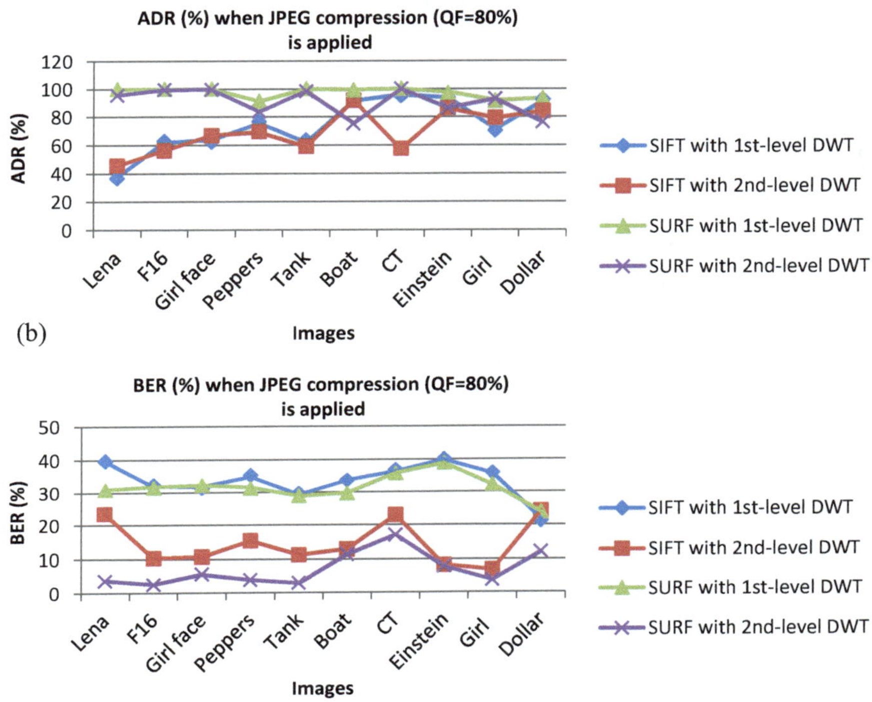

(b)

Fig. 3.16 Performance comparison in terms of (**a**) ADR and (**b**) BER when JPEG compression (QF = 80%) is applied

respectively. These measurements are comparatively inadequate, since the core target of the present work is to achieve a reliable and robust stego-system with high imperceptibility. However, the conflicting parameters of the steganography systems (imperceptibility, robustness, and payload capacity) are at odds with each other. Therefore, the CR-BIS system efficiently meets robustness and imperceptibility in tradeoff with the hiding capacity. The limited hiding capacity could be improved either by increasing the number of the characteristic regions used for embedding or by choosing smaller thresholds for SURF. Alternatively, with smaller thresholds, the SURF technique detects the embedding regions with less accuracy, which negatively affects the reliability of the stego-system after undergoing attacks. Moreover, the techniques that attempt to hide large amounts of information may alter the statistical properties of the image and make the technique easily detectable (Kruus, Scace, Heyman, & Mundy, 2003).

Comparing the proposed CR-BIS algorithm with Li's algorithm, the experimental results in Tables 3.1, 3.2, 3.3, 3.4, 3.5, 3.6, 3.7, 3.8, 3.9, 3.10, and 3.11 and Figs. 3.13, 3.14, 3.15, 3.16, 3.17, 3.18. 3.19, 3.20, 3.21, 3.22, 3.23, 3.24, and 3.25

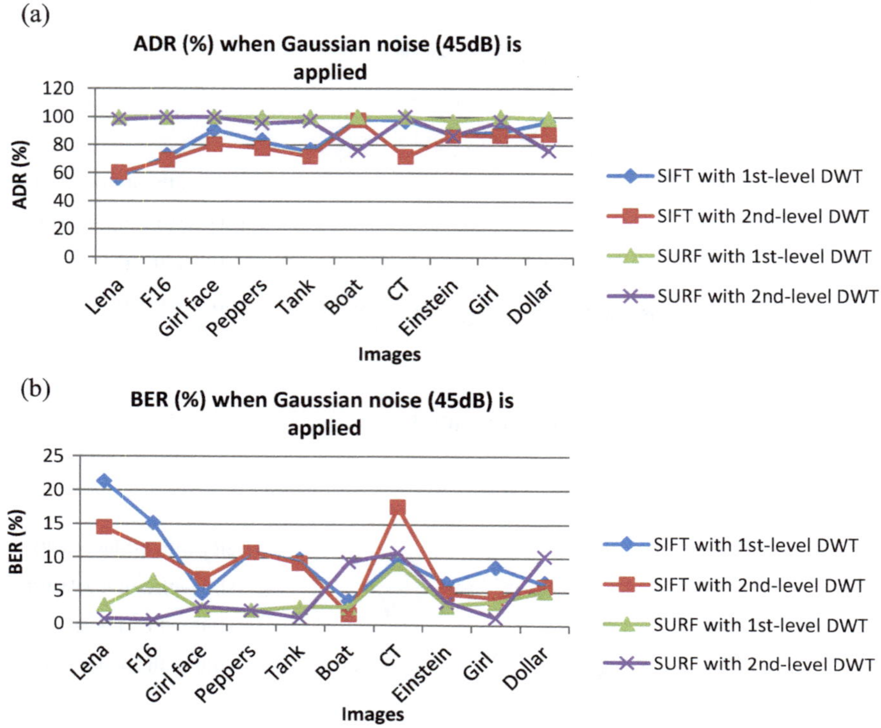

Fig. 3.17 Performance comparison in terms of (**a**) ADR and (**b**) BER when Gaussian noise (45 dB) is applied

demonstrate the advantages of using SURF to show the higher robustness level indicated by the lower BER values (Hamid et al., 2012a, 2013a). The proposed algorithm can achieve higher ADR values than Li's algorithm, as shown in Fig. 3.25a. The higher ADR values obtained in this present study are attributed to the implementation of the most appropriate algorithm—SURF—that meets the requirements of a high security system. The experimental results proved that SURF can detect the characteristic regions efficiently during the extraction phase, although the image undergoes attacks. The robustness of the SURF-based scheme increases when second-level DWT is used to hide data, especially against JPEG compression. Utilizing higher DWT levels is useful to enhance robustness. However, the method negatively affects visual quality in terms of PSNR, as shown in Table 3.12. Furthermore, the higher DWT levels affect the abilities of SURF and SIFT to extract the interest points correctly, because higher DWT levels result in developed image degradation. However, the visual superiority of the stego-images remains high, as the PSNR values are within a suitable choice (33.53–45.98 dB).

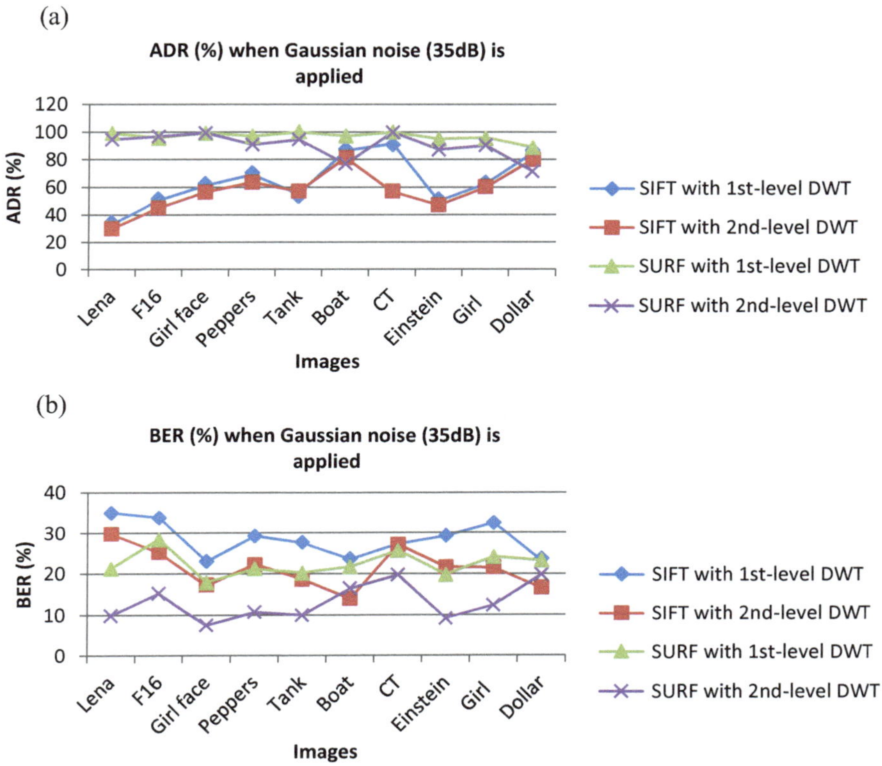

Fig. 3.18 Performance comparison in terms of (**a**) ADR and (**b**) BER when Gaussian noise (35 dB) is applied

The benefits of the proposed scheme can be summarized as follows:

- The security of the embedded message is enhanced by exploiting a powerful encryption technique along with dynamic regions of embedding by using SURF.
- It has high imperceptibility in terms of high PSNR values.
- It possesses certain levels of robustness to signal processing attacks.

However, the algorithm has the following limitations:

- The embedding capacity is limited.
- It is not robust to all types of attacks.

The imperceptibility and security are the main concerns of the present research. Therefore, the disadvantages can be overcome or manipulated. If a single image is not enough to hide the secret message, then the secret information could be distributed over several images to overcome the capacity-limitation issue (Hamid et al., 2012b, 2013b). Moreover, the robustness of the secret data can be enhanced by duplicating the message bits or by employing error checking and correction techniques, which will be discussed in Chap. 4.

(a)

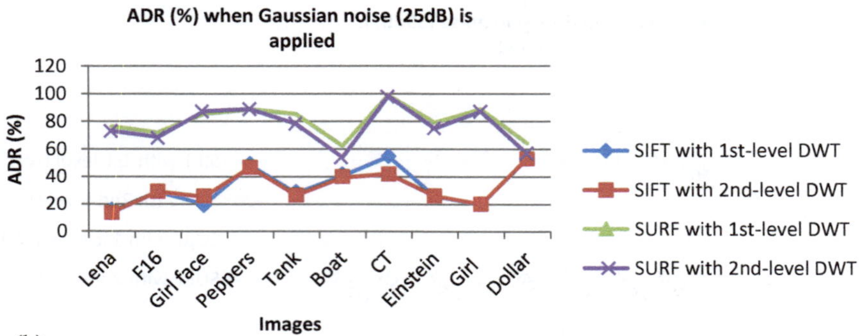

(b)

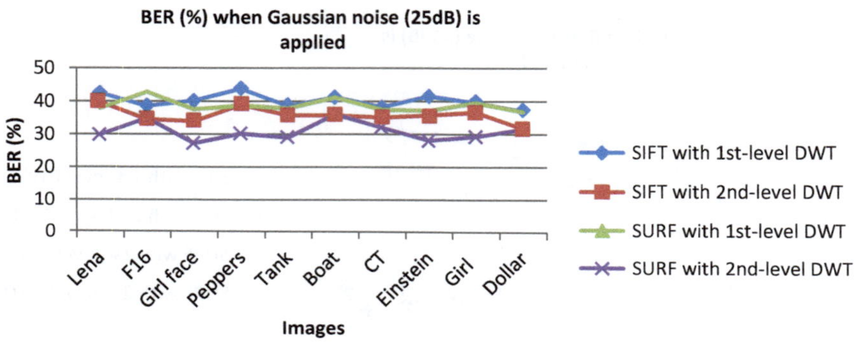

Fig. 3.19 Performance comparison in terms of (**a**) ADR and (**b**) BER when Gaussian noise (25 dB) is applied

3.7 Summary

For most of the current steganography techniques, the information-hiding process modifies almost all cover components. Hiding data in the whole image may affect visual quality and increases the possibility of data loss after any possible attacks. In this chapter, a new region-based steganography method, CR-BIS, which hides data in the robust regions of the image, is proposed. First, the secret data are encrypted via a highly secure encryption algorithm. Second, SURF is used to locate the strongest sections in the image. Then data embedding is accomplished in a content-based style by varying the wavelet transform coefficients of those strong sections.

The robustness of the proposed algorithm increases when second-level DWT is used to hide data, especially against JPEG compression. However, applying the same scheme to the median and the low-pass filters remains difficult. Utilizing

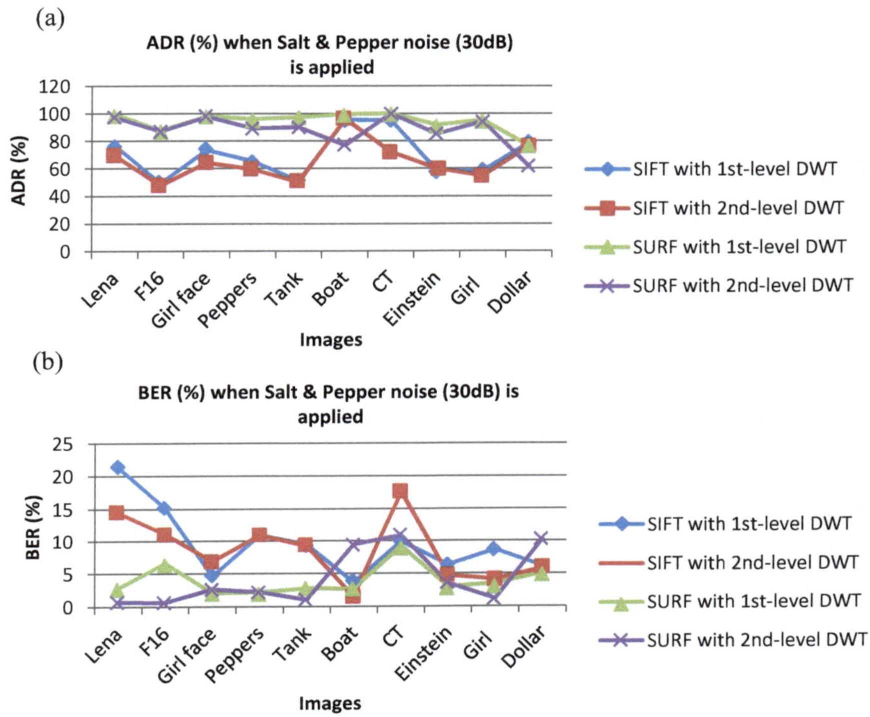

Fig. 3.20 Performance comparison in terms of (**a**) ADR and (**b**) BER when salt and pepper noise (30 dB) is applied

higher DWT levels is useful to enhance the robustness. However, the method has a negative effect on the visual quality in terms of PSNR. Moreover, higher DWT levels affect the abilities of SURF and SIFT to remove the main points correctly because higher DWT levels result in higher image degradation. Yet the visual quality of the stego-images remains high as the PSNR values are within an acceptable range. The achieved hiding capacity is relatively limited, which makes the scheme more appropriate for copyright protection applications. To use this algorithm to transmit a large amount of secret data, the data could be divided among several images.

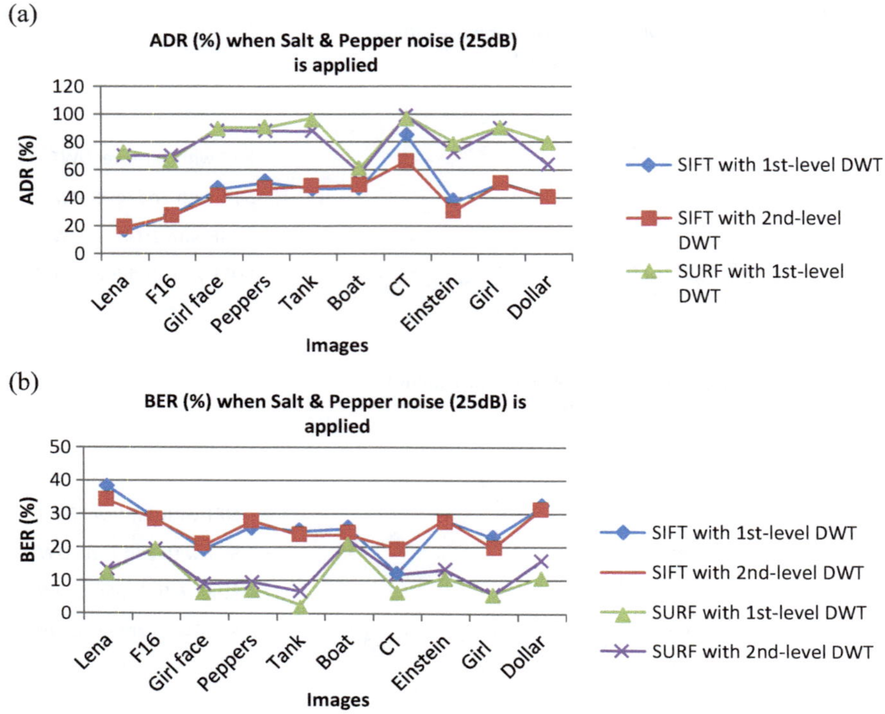

Fig. 3.21 Performance comparison in terms of (**a**) ADR and (**b**) BER when salt and pepper noise (25 dB) is applied

(a)

(b)

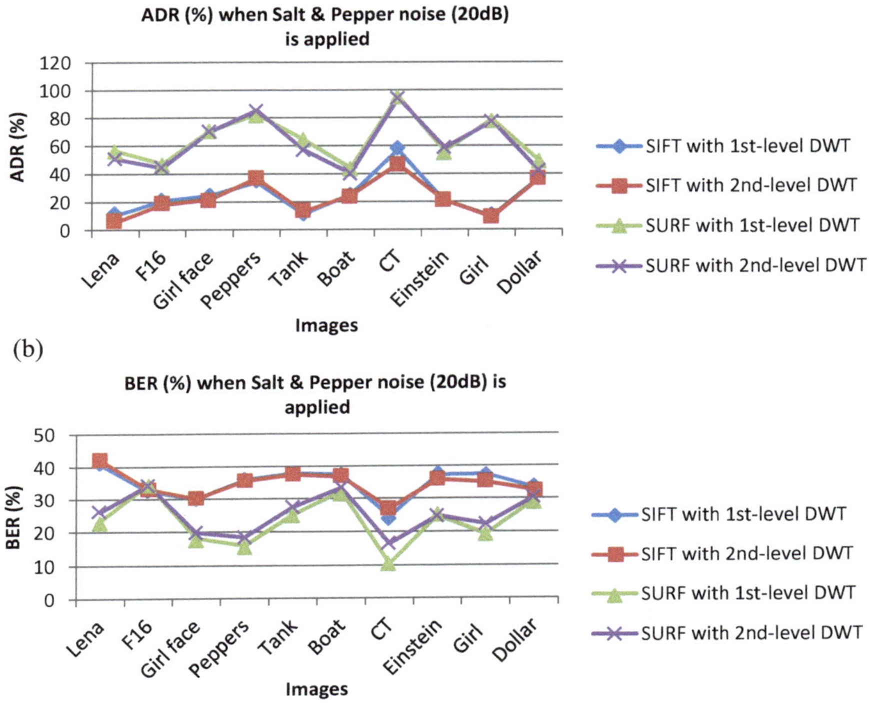

Fig. 3.22 Performance comparison in terms of (**a**) ADR and (**b**) BER when salt and pepper noise (20 dB) is applied

(a)

(b)

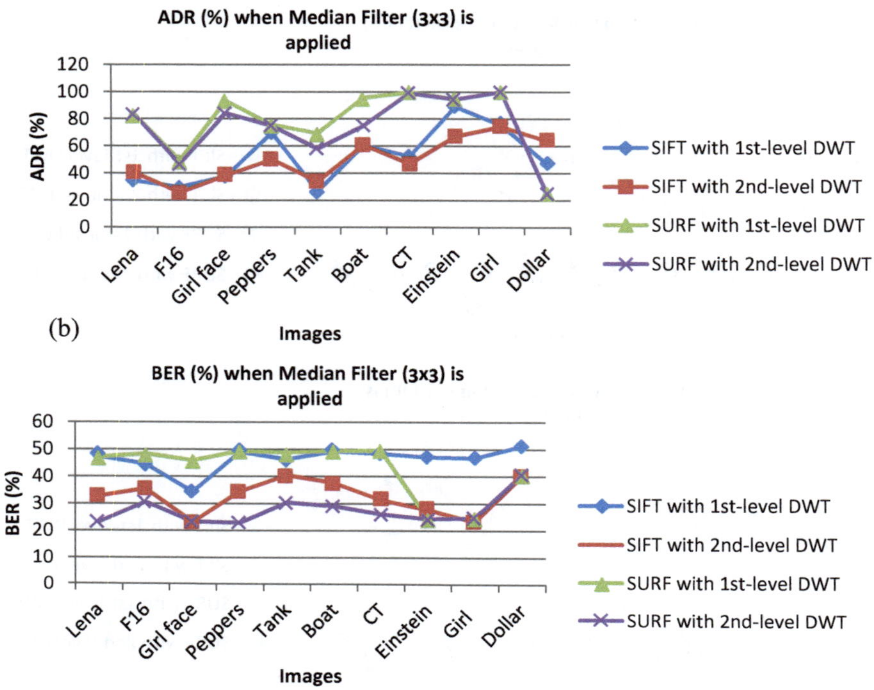

Fig. 3.23 Performance comparison in terms of (**a**) ADR and (**b**) BER when median filter (3 × 3) is applied

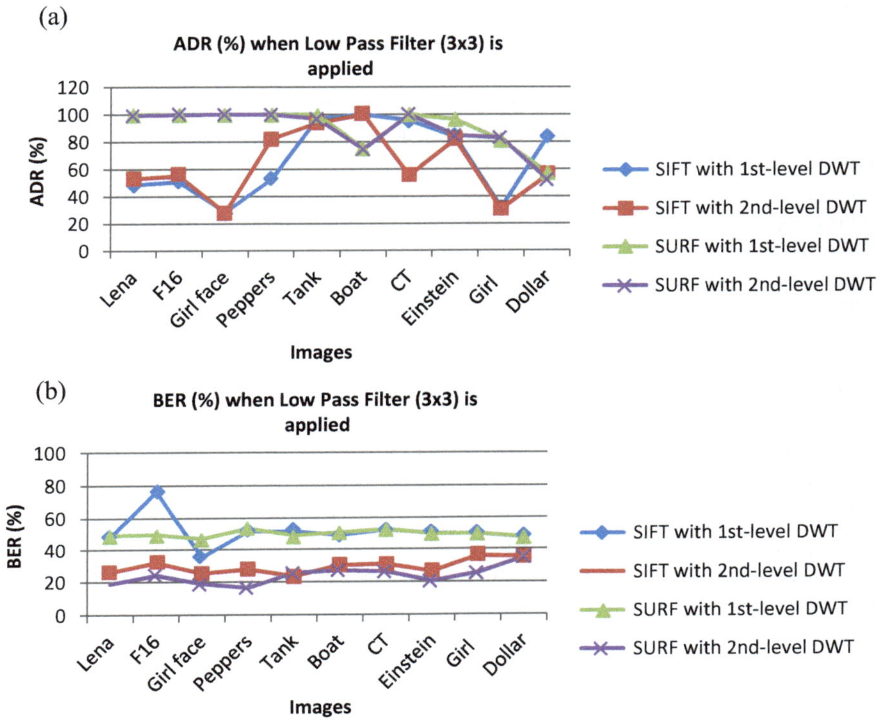

Fig. 3.24 Performance comparison in terms of (**a**) ADR and (**b**) BER when low pass filter (3 × 3) is applied

Table 3.11 A comparison between Li's algorithm and the CR-BIS algorithm in terms of the average of ADR and BER for all images used for testing

	Li' algorithm				The proposed algorithm			
	1-level DWT		2-level DWT		1-level DWT		2-level DWT	
Type of attack	ADR (%)	BER (%)	ADR (%)	BER (%)	ADR (%)	BER (%)	ADR (%)	BER (%)
No attack	94.45	2.89	87.79	4.93	99.72	0.79	93.06	3.96
JPEG compression (QF = 100%)	93.28	3.68	86.61	5.39	99.65	0.95	92.87	4.04
JPEG compression (QF = 90%)	80.74	33.45	73.89	14.61	99.16	31.53	92.75	7.12
JPEG compression (QF = 80%)	74.49	40.42	69.54	20.59	97.24	39.90	90.79	17.13
AWGN (45 dB)	84.72	9.61	79.05	8.63	99.60	4.00	92.58	4.26
AWGN (35 dB)	64.51	28.43	57.74	21.37	96.78	22.35	90.19	13.12
AWGN (25 dB)	33.36	40.07	32.13	35.75	80.09	38.65	76.69	30.79
Salt & pepper noise (30 dB)	70.19	13.55	64.96	14.89	94.12	4.33	87.90	7.12
Salt & pepper noise (25 dB)	44.72	25.71	41.92	25.65	83.11	10.61	78.82	12.72
Salt & pepper noise (20 dB)	25.03	34.69	23.22	34.51	64.55	23.29	62.05	25.37
Median filter (3 × 3)	52.28	46.39	50.15	32.61	78.64	42.79	74.04	27.52
Low pass filter (3 × 3)	67.20	51.04	63.48	29.21	90.98	49.70	89.00	23.72

(a)

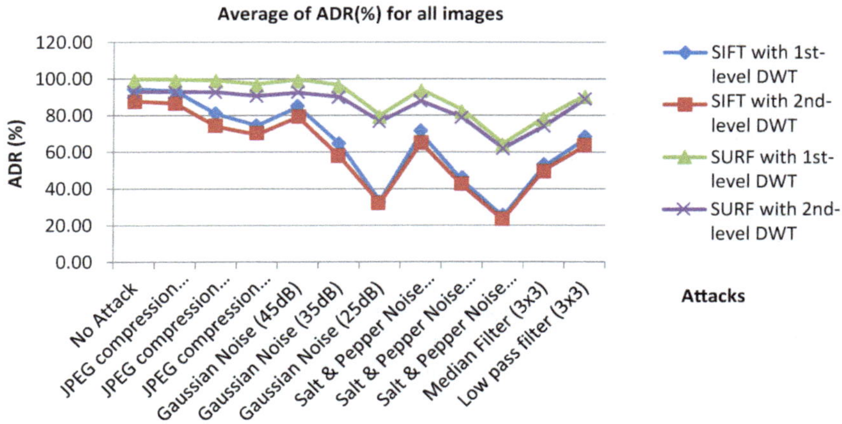

(b)

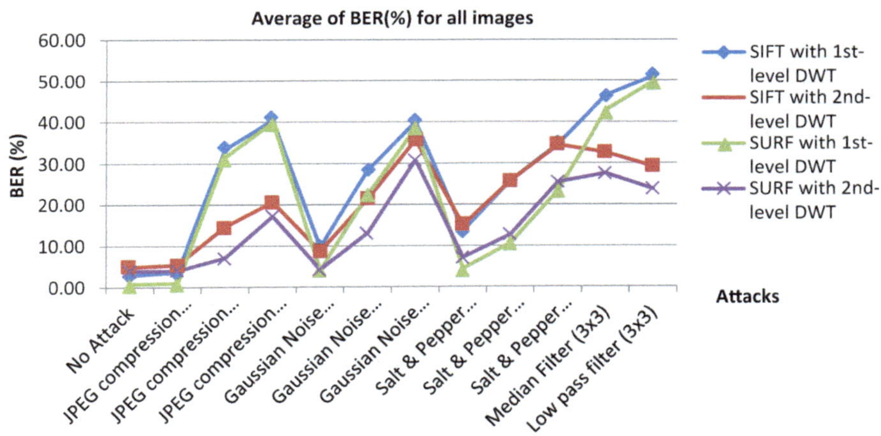

Fig. 3.25 Performance measured by average of (**a**) ADR and (**b**) BER for all images used for testing after applying different types of attacks

Table 3.12 Hiding capacity and the corresponding PSNR achieved using the CR-BIS algorithm

	1st-level DWT		2nd-level DWT	
Test image	Hiding capacity (bits)	PSNR (dB)	Hiding capacity (bits)	PSNR (dB)
Lena	15,696	45.38	3096	42.14
F16	10,464	45.91	2064	41.25
Girl face	15,696	45.30	3096	43.15
Peppers	10,464	48.30	2064	47.34
Tank	10,464	45.98	2064	45.18
Boat	10,464	43.84	2064	41.54
CT	15,696	42.25	3096	41.23
Einstein	13,080	45.69	2580	45.33
Girl	10,464	47.67	2064	45.45
Dollar	15,696	33.53	3096	34.96

References

Abdul-mahdi, N. H., Yahya, A., Ahmad, R. B., & Al-Qershi, O. M. (2013). Secured and robust information hiding scheme. *Procedia Engineering Journal, 53*, 463–471.

Bauer, J., Sünderhauf, N., & Protzel, P. (2006). *Comparing several implementations of two recently published feature detectors*. Paper presented at the International Conference on Intelligent Autonomous Vehicles, Toulouse, France.

Bay, H. (2006). *From wide-baseline point and line correspondences to 3D*. PhD Doctoral and Habilitation Thesis Swiss Federal Institute of Technology, ETH Zurich, Zürich.

Cover, K. S. (2006). Multiexponential reconstruction algorithm immune to false positive peak detection. *Review of Scientific Instruments, 77*(7), 075101–075115.

Cheddad, A., Condell, J., Curran, K., & Mc Kevitt, P. (2009). A skin tone detection algorithm for an adaptive approach to steganography. *Signal Processing, 89*(12), 2465–2478.

Cheddad, A., Condell, J., Curran, K., & McKevitt, P. (2008a, March 31–April 4). *Biometric inspired digital image steganography*. Paper presented at the Engineering of Computer Based Systems, 2008. ECBS 2008. 15th Annual IEEE International Conference and Workshop on the.

Cheddad, A., Condell, J., Curran, K., & McKevitt, P. (2008b, May 28–30). *Enhancing steganography in digital images*. Paper presented at the Computer and Robot Vision, 2008. CRV '08. Canadian Conference on.

Cheddad, A., Condell, J., Curran, K., & Kevitt, P. M. (2010). Digital image steganography: Survey and analysis of current methods. *Signal Processing, 90*(3), 727–752.

Daemen, J., & Rijmen, V. (2002). *The design of Rijndael: AES - the advanced encryption standard; with 17 tables*. Berlin [u.a.]: Springer.

Finch, P. J. M. (1995). *A study of the Blowfish encryption algorithm*. New York: City University of New York.

Hamid, N., Yahya, A., Ahmad, R. B., & Al-Qershi, O. M. (2012a). Image steganography techniques: An overview. *International Journal of Computer Science and Security (IJCSS), 6*(3), 168–178.

Hamid, N., Yahya, A., Ahmad, R. B., & Al-Qershi, O. M. (2012b, April 10–12). *Characteristic region based image steganography using speeded-up robust features technique*. Paper presented at the 1st International Conference on Future Communication Network (ICFCN'12). IEEE International Conference, Iraq, Baghdad.

Hamid, N., Yahya, A., Ahmad, R. B., Najim, D., & Kanaan, L. (2013a). Steganography in image files: A survey. *Australian Journal of Basic and Applied Sciences, 7*(1), 35–55.

Hamid, N., Yahya, A., Ahmad, R. B., Najim, D., & Kanaan, L. (2013b). Enhancing the robustness of digital image steganography using ECC and redundancy. *WULFENIA Journal, 20*(4), 153–169.

Juan, L., & Gwun, O. (2009). A comparison of SIFT, PCA-SIFT and SURF. *International Journal of Image Processing (IJIP), 3*(4), 143–152.

Karthigai Kumar, P., & Baskaran, K. (2010). An ASIC implementation of low power and high throughput blowfish crypto algorithm. *Microelectronics Journal, 41*(6), 347–355.

Koenderink, J. J. (1984). The structure of images. *Biological Cybernetics, 50*, 363–370.

Kruus, P., Scace, C., Heyman, M., & Mundy, M. (2003). A survey of steganographic techniques for image files. *Advanced Security Research Journal, V(I)*, 41–52.

Kumar, M. A., & Karthikeyan, S. (2012). Investigating the efficiency of Blowfish and Rejindael (AES) algorithms. *I. J. Computer Network and Information Security, 4*(2), 22–28.

Kumar, V., & Kumar, D. (2010). In V. V. Das & R. Vijaykumar (Eds.), *Digital image steganography based on combination of DCT and DWT information and communication technologies* (Vol. 101, pp. 596–601). Berlin Heidelberg: Springer.

Lindeberg, T. (1990). Scale-space for discrete signals. *IEEE Transactions on Pattern Analysis and Machine Intelligence, 12*(3), 234–254.

Li, L., Qian, J., & Pan, J. S. (2011). Characteristic region based watermark embedding with RST invariance and high capacity. *AEU - International Journal of Electronics and Communications, 65*(5), 435–442.

Li, Y., Li, C.-T., & Wei, C.-H. (2007). *Protection of mammograms using blind steganography and watermarking*. Paper presented at the Proceedings of the Third International Symposium on Information Assurance and Security.

Lowe, D. G. (2004). Distinctive image features from scale-invariant keypoints. *International Journal of Computer Vision, 60*(2), 91–110.

Ming-Kuei, H. (1962). Visual pattern recognition by moment invariants. *IRE Transactions on Information Theory, 8*(2), 179–187.

Milad, A. A., Muda, H. Z., Noh, Z. A. B. M., & Algaet, M. A. (2012). Comparative study of performance in cryptography algorithms (Blowfish and Skipjack). *Journal of Computer Science, 8*(7), 1191–1197.

Potdar, V. M., Han, S., & Chang, E. (2005, August 10–12). *A survey of digital image watermarking techniques*. Paper presented at the Industrial Informatics, 2005. INDIN '05. 2005 3rd IEEE International Conference on.

Rijmen, V. (1997). *Cryptanalysis and design of iterated block ciphers*. Ph.D. Dissertation, Katholieke Universiteit Leuven, Belgian.

Schneier, B. (1994). *Description of a newvariable-length key, 64-bit block cipher (Blowfish) Lecture notes in computer science* (Vol. 809). Berlin, Heidelberg: Springer.

Schneier, B. (1995). *Applied cryptography: Protocols, algorithms, and source code in C*. John Wiley & Sons, Inc.

Schneier, B. (2012). *Liars and outliers: enabling the trust that society needs to thrive*. ISBN: 978-1-118-14330-8. Indianapolis, IN: John Wiley & Sons, Inc.

Singh, G., Singla, A. K., & Sandha, K. S. (2012). Superiority of blowfish algorithm in wireless networks. *International Journal of Computer Applications, 44*(11), 23–26.

Silva, E. A., & Agaian, S. S. (2004). *The best transform in the replacement coefficients and the size of the payload relationship sense*. Paper presented at the Society for Imaging Science & Technology, 2004, USA.

Teague, M. R. (1980). Image analysis via the general theory of moments*. *Journal of the Optical Society of America, 70*(8), 920–930.

Tingyuan, N., Chuanwang, S., & Xulong, Z. (2010, April 23–25). *Performance evaluation of DES and Blowfish algorithms*. Paper presented at the Biomedical Engineering and Computer Science (ICBECS), 2010 International Conference on.

Verma, H. K., & Singh, R. K. (2012). Performance analysis of RC5, Blowfish and DES block cipher algorithms. *International Journal of Computer Applications, 42*(16), 8–14.

Chapter 4
An Enhanced Robust and Protected Image Steganographic System

Abstract This chapter deals with two main aspects; the first one is concerned with introducing an improved robust and secure image steganography system. The original algorithm presented by Mali et al. gives adequate results in terms of robustness; however, the algorithm cannot be considered reliable, as some data is lost. In order to overcome the problem of the lost data due to the misidentified blocks, an embedding map was proposed to specify the location of the blocks that can be used for embedding a secret message. Such a technique, the embedding map, is very important if one needs to start the extracting phase correctly and accurately. Besides, any loss in the embedding map will in turn lead to data loss. Consequently, the embedding map will be hidden using SURF and DWT to assure robustness and to increase the level of security of the proposed system. The secret data can be embedded using the original algorithm (DCT-based) proposed by Mali et al. More so, the experimental results show the ability of the new algorithm, namely IRSS, to overcome the problem of losing data even with JPEG compression or Gaussian noise.

4.1 Introduction

This chapter discusses in detail the design of the IRSS algorithm. Such an algorithm exploits the main benefits of the CR-BIS algorithm which was introduced in Chap. 3. A developed algorithm is proposed to successfully embed secret data within the frequency domain by modifying the DCT coefficients. Based on Image Adaptive Energy Thresholding (AET), certain blocks are selected for the concealment of secret data (Hamid et al., 2012a,b). To ensure a full recovery for the hidden message, an embedding map is proposed to indicate the selected embedding blocks. To secure the embedding map, the SURF technique is used to dynamically define the robust locations of the image where the embedding map is to be concealed. In addition, the embedding map is hidden in the frequency domain as well, by modifying the DWT coefficients in a content-based manner. The analysis of the proposed scheme is evaluated by considering the robustness against AWGN and JPEG compression attacks. Moreover, the resultant stego-images demonstrate a good visual

A. Yahya, *Steganography Techniques for Digital Images*,
https://doi.org/10.1007/978-3-319-78597-4_4

quality in terms of PSNR. Nevertheless, due to the fact that specific parts of the image serve to hide the embedding map, the hiding capacity achieved in this work is somewhat limited (Hamid et al., 2013a,b).

Recently, (Mali, Patil, & Jalnekar, 2012) proposed a robust DCT-based steganographic scheme via a powerful coding framework that allows the dynamic choice of hiding locations and the embedding of low and medium DCT coefficients. The robustness of their scheme not only comes from exploiting low and medium DCT coefficients for hiding data, but also mainly from the redundancy, whereby the payload bits are repeated (n) times in order to add robustness to the system. However, the scheme has a severe drawback that results in a loss of information. In this chapter, the original scheme proposed by (Mali et al., 2012) and its drawbacks are presented in Sect. 4.2.1. In Sect. 4.2.2, a proper modification is proposed to overcome the adopted scheme drawbacks and make it more applicable by inserting the embedding map. Moreover, this chapter discusses the effect of using ECC and redundancy on enhancing the robustness of digital image steganography as detailed in Sect. 4.3. Section 4.3.1 explains the main concepts of the DWT quantization method. Section 4.3.2 discusses the principles of the histogram shifting technique. The fundamental concepts of the RS-code have been clarified in Sect. 4.3.3. In Sect. 4.3.4, another technique briefly expounded for completeness is the DCT-based quantization technique (Hamid et al., 2012a, 2013a).

4.2 DCT-Based Image Steganography

DCT has been widely used for steganography and watermarking purposes. The DCT-based methods hide data bits in significant areas of the cover-image in order to make them more robust to attacks. Generally, DCT is applied to image blocks of 8×8 pixels, and selected coefficients, of some selected blocks, are used to hide data bits. The coefficients are modified differently in order to reflect an embedding of "0" or "1." More details about steganography in the DCT domain were explained in Chap. 2, Sect. 2.4.2.

In 2012, (Mali et al., 2012) proposed a steganographic method for embedding a high volume of text information in digital cover-images without leaving perceptual distortion. Figure 4.1 shows the image steganographic system proposed by (Mali et al., 2012). Mali et al.'s scheme consists of two main stages: processing the data to be embedded, and embedding the data. In the first stage, the pure payload bits undergo three processes:

1. Encryption: to secure the data;
2. Redundancy addition: to reduce the BER; and
3. Interleaving: to ensure that the redundant bits are spread all over the image.

Both redundancy and interleaving are responsible for the recovery of the robust data at the receiver end. Nevertheless, the overall robustness also depends on the

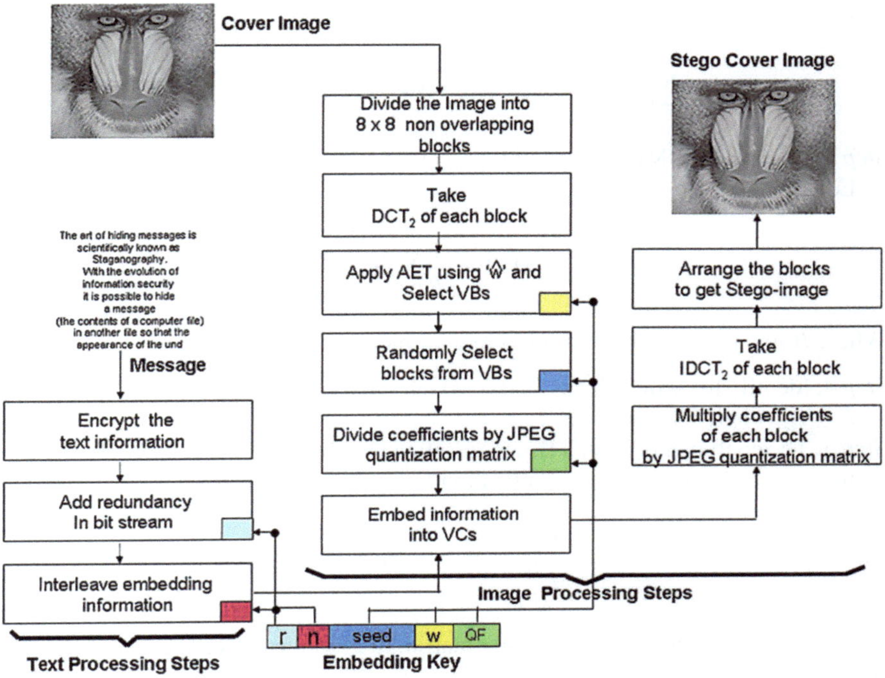

Fig. 4.1 General steganographic system proposed by (Mali et al., 2012)

embedding procedures. The embedding procedures and their drawbacks are discussed in the following section.

4.2.1 Mali's Embedding Procedures

After processing the data to be embedded, the inputs to the embedding system are a cover-image file (C), the processed text (FBS), Energy Threshold Factor ($w^\wedge$), and JPEG quality factor (QF). The embedding phase can be summarized in this section by the following steps:

Step 1: Divide the image into 8 × 8 non-overlapping blocks, so that if the intensity values of the 8 × 8 blocks are a_{ij}, then the corresponding DCT coefficients C_{ij} are given by:

$$C_{ij} = \mathrm{DCT}_2\left(a_{ij}\right) \tag{4.1}$$

Where $i, j = \{0, 1, 2, \ldots\ldots\ldots, 7\}$ and DCT_2 denotes a two-dimensional DCT.

Step 2: Calculate the Energy of each block as:

$$E = \sum_{i=1}^{7} \sum_{j=1}^{7} \left\| C_{ij} \right\|^2 \ \forall i, j = \{0,1,2,\ldots\ldots,7\}, (i,j) \neq 0 \tag{4.2}$$

Step 3: Calculate the Mean Value of Energy (MVE) of the image using the following Equation:

$$\text{MVE} = \frac{1}{B} \sum_{1}^{B} E_b \tag{4.3}$$

Where B=the total number of blocks and b= block number.

Step 4: Identify the Valid Blocks (VBs), which satisfy the Energy Threshold Criteria, $E \geq E_T$, where $E_T = w^{\wedge} \times \text{MVE}$.

Step 5: The coefficients of all VBs are quantized by dividing them according to their respective elements of the quantization matrix, as stated below:

$$C_{ij}{}^{\wedge} = \frac{C_{ij}}{M_{ij}^{QF}} \quad \forall i, j = \{0,1,\ldots\ldots\ldots,7\} \tag{4.4}$$

Where $C_{ij}{}^{\wedge}$ is the quantized coefficient matrix, $M_{ij}{}^{\text{QF}}$ is the *ijth* element of the quantization matrix for a given value of QF; as for QF, it represents a parameter that controls the image quality and the extent to which the image can be compressed. That is, a higher value of QF corresponds to a higher-quality image and a larger file size. On the other hand, the smaller the value of QF, the more the loss of information and the smaller the file size (Malik & Baharudin, 2013).

Step 6: Identify the Valid DCT Coefficients (VCs) which satisfy the non-zero criteria ($C_{ij} \neq 0$) and fall into the lower and middle frequency band.

Step 7: The coefficients of all VCs are scanned in zigzag fashion to get the one dimensional vector $C_k{}^{\wedge}$. The process of embedding data makes the quantized non-zero DCT coefficients *Odd* for '*bit* = 0' or *Even* for '*bit* = 1'. The coefficients with hidden bits $d_k{}^{\wedge}$ are given by:

$$d_k{}^{\wedge} = \begin{cases} \text{Odd}\, C_k{}^{\wedge}, \text{if bit} = 0 \\ \text{Even}\, C_k{}^{\wedge}, \text{if bit} = 1 \end{cases} \tag{4.5}$$

Step 8: The hidden coefficients $d_k{}^{\wedge}$ are reversely scanned to form an 8 × 8 matrix. It is then multiplied by the JPEG quantization matrix to obtain unquantified coefficients C_{ij}.

Step 9: Apply inverse DCT to each block, and reconstruct the image as a stego-image.

From the above steps, it is clear that the extraction phase depends mainly on identifying the blocks that have been used for the correct embedding. Misidentification

71	68	75	85	75	75	67	81
71	67	75	80	73	74	77	88
77	78	75	79	81	80	90	100
78	74	75	69	82	92	100	112
80	75	85	81	88	99	109	121
80	81	84	88	100	112	122	133
77	81	95	105	113	121	135	141
81	91	105	114	122	131	143	144

$MVE = 565{,}670$ and $E = 568{,}185$

$w^{\wedge} = 1 \rightarrow E > E_T$

(a)

71	70	83	85	79	80	58	83
66	61	75	73	70	77	70	92
73	68	72	69	73	81	85	103
81	66	76	65	76	94	97	111
85	65	89	84	84	104	109	116
83	65	86	92	96	119	125	127
86	66	97	109	105	126	140	134
100	82	111	119	112	135	148	135

$MVE = 566{,}780$ and $E = 560{,}244$

$w^{\wedge} = 1 \rightarrow E < E_T$

(b)

Fig. 4.2 Example of an undetected block because $E <$ MVE (**a**) The original block (**b**) After embedding

of those blocks will cause the loss of a portion of the embedded data or extracting the unwanted data. During the extracting phase, the blocks that have been used for embedding data should be identified first. Such a process is achieved by the following two steps:

First: The Energy of the block E should be $\geq E_T$ (Step 4).

Second: The lower and middle DCT coefficients of the block should satisfy the non-zero criteria (Step 6). If the algorithm fails to identify the blocks in any of these two conditions, this will lead to incorrect extraction of the embedded data.

The algorithm has been implemented and simulated through the use of different images and randomly generated data. The obtained results show that the algorithm may misidentify the blocks in some cases. For example, in one of the experiments on the standard image "Lena," it has been noted that the embedding process results in changing the MVE and E values. Accordingly, the embedded blocks could not be identified during the extraction phase because ($E < E_T$). In another situation, reconstructing the stego-image involves mathematical rounding operations to get the integer pixel values. The rounding operations may turn some non-zero coefficients to zero coefficients or vice versa. It may further cause misidentification to the blocks which carry the data. Figure 4.2 shows an example of one of the undetected blocks in the standard image "Lena," due to the aforementioned reasons.

In view of that, Table 4.1 illustrates the number of the misidentified blocks after applying Mali et al.'s algorithm on "Lena's" image, using different QF values and different energy threshold values ($w^{\wedge}$). Such a problem severely affects the integrity

Table 4.1 Misidentified blocks after applying Mali et al.'s algorithm on the image of "Lena"

Energy thresholding $w^{\wedge}$=	Number of misidentified blocks			Percentage of misidentified blocks		
	QF = 50%	QF = 75%	QF = 100%	QF = 50%	QF = 75%	QF = 100%
0.0.5	35	27	51	1.17	0.90	1.71
0.6	30	28	47	1.11	1.03	1.73
0.7	35	27	39	1.39	1.07	1.55
0.8	21	23	44	0.90	0.98	1.88
0.9	42	14	39	2.00	0.67	1.85
1	18	21	40	0.98	1.15	2.19

of the embedded data and the reliability of the stego-system; especially, when one cannot identify which blocks have been misidentified. Hence, Mali et al.'s scheme cannot be considered a reliable and applicable steganographic system. This is because the most important parameter of the applicable steganographic system, in addition to robustness, imperceptibility, and payload capacity, is the reliability. The latter indicates that the embedded data should be successfully retrieved (with 100% recovery).

To solve the problem of block misidentification, an embedding map (location map) is proposed in this regard. The concept of the embedding map has been used in some data-hiding techniques to correctly identify the location of the blocks or regions where data has been embedded (Alattar, 2004; Jun, 2003). That is to say, it is not the whole image that is used for the process of embedding. In the following section, Mali et al.'s algorithm is modified by incorporating an embedding map.

4.2.2 Modifying Mali et al.'s Scheme

The proposed algorithm (IRSS) is mainly based on (Mali et al., 2012). In order to overcome the problem of block misidentification, an embedding map technique is introduced in this work to ensure a full recovery for the hidden information (with 100% recovery) (Hamid et al., 2013a,b). Exploiting an embedding map implies generating a binary map of a size that is equal to the number of blocks in the image. If the image size is $(m \times n)$, then, the embedding map size should be $(\frac{m}{8} \times \frac{n}{8})$, where the block size is (8 × 8) pixels. Each block in the image is represented by a bit in the embedding map. Besides, if the bit is '1,' this means that the corresponding block is used for embedding the information and vice versa. Figure 4.3 illustrates the process of generating the binary embedding map. In such a mapping process, the selected blocks for embedding are denoted by (Yes), and those blocks which are not valid for data embedding are indicated by (No). The embedding map is to be concealed in the robust regions of the image that are identified by implementing the SURF technique. At the same time, the secret data is concealed in other regions or blocks. Such regions, in which the data are hidden,

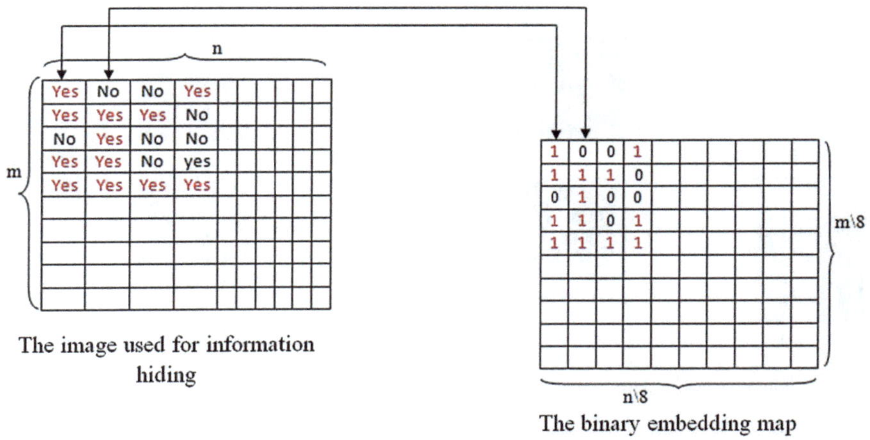

Fig. 4.3 Generating the binary embedding map

are determined according to Mali et al.'s method. The embedding map could be embedded in some predefined regions selected by the user. However, this option affects the data security, because the same regions are used every time. Therefore, it is important to hide the embedding map in a confidential and secure manner, as discussed and explained in the next subsection.

4.2.3 Hiding the Embedding Map

The embedding map is necessary to initiate the extracting phase, which means it must be concealed in the image in an extremely secure and robust way (Abdul-mahdi et al., 2013). To achieve these two objectives, two powerful techniques have been used.

The first technique is used to guarantee the security of the embedding map. It is performed by selecting the most appropriate regions for embedding in a dynamic way, depending on the key-points (interest points) of the image. For this purpose, the SURF technique is exploited to extract the distinctive local features in the image and to produce the interest-point descriptors that demonstrate those features. Each feature vector has some information that describes its corresponding interest point. However, as described in Chap. 3, Sect. 3.3.2, the SURF technique is used to calculate a list of invariant interest points in the cover image. The detected interest points will be the centers of the regions in which the embedding map is to be concealed. Consequently, twelve non-overlapping interest points with the highest scales are selected and used to hide the embedding map. These interest points are the most robust features in the image and can be detected even when the stego-image undergoes different types of operations, such as that of JPEG compression and Gaussian

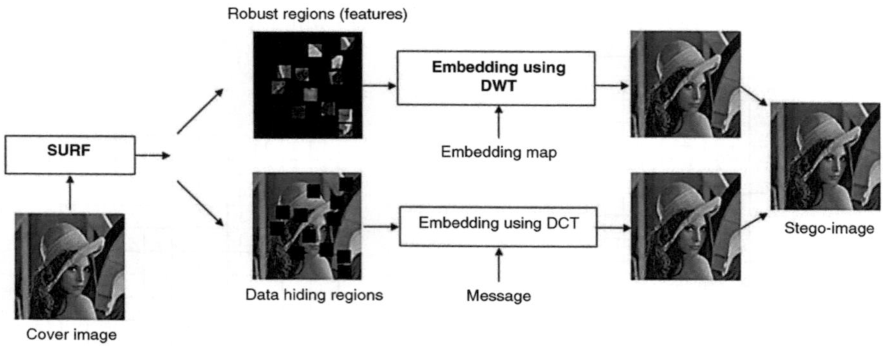

Fig. 4.4 The process of data embedding using the IRSS algorithm

noise. As to the second requirement, robustness, a DWT-based embedding technique is adopted in which the data are embedded in a content-based manner, as clarified in Chap. 3, Sect. 3.3.3. The proposed algorithm reported in this section can be described through the following steps (Hamid et al., 2012a,b):

Step 1: The SURF technique is applied to the image, and 12 invariant interest points with the highest scales are identified. The center that coordinates these interest points defines 12 square regions of (32 × 32) sized pixels. These regions are used to hide the embedding map.

Step 2: The image is then divided into (8 × 8) non-overlapping blocks so that DCT is applied to each block a_{ij} to get C_{ij}, as given in Eq. (4.1). It should be taken into consideration that the blocks that intersect with 12 square regions, as outlined in Step 1, are discarded, since those regions are used to hide the embedding map, not the data.

Step 3: The blocks are scanned, and the blocks which can be used for embedding are identified (according to Mali et al.'s algorithm), as described in Sect. 4.2.1.

Step 4: An embedding map is constructed to indicate the blocks in which the data bits will be embedded.

Step 5: The secret data are embedded in the blocks defined in Step 3 by modifying the DCT coefficients according to Mali et al.'s algorithm (Mali et al., 2012).

Step 6: The embedding map is embedded frequently in the 12 square regions defined in Step 1 by modifying the DWT coefficients (in a content-based manner). The addition of the redundant bits and interleaving techniques has been used to prepare both data and the embedding map to be hidden. Figure 4.4 illustrates the process of data embedding, and Fig. 4.5 showcases the flowchart of the currently proposed robust and secured stego-system.

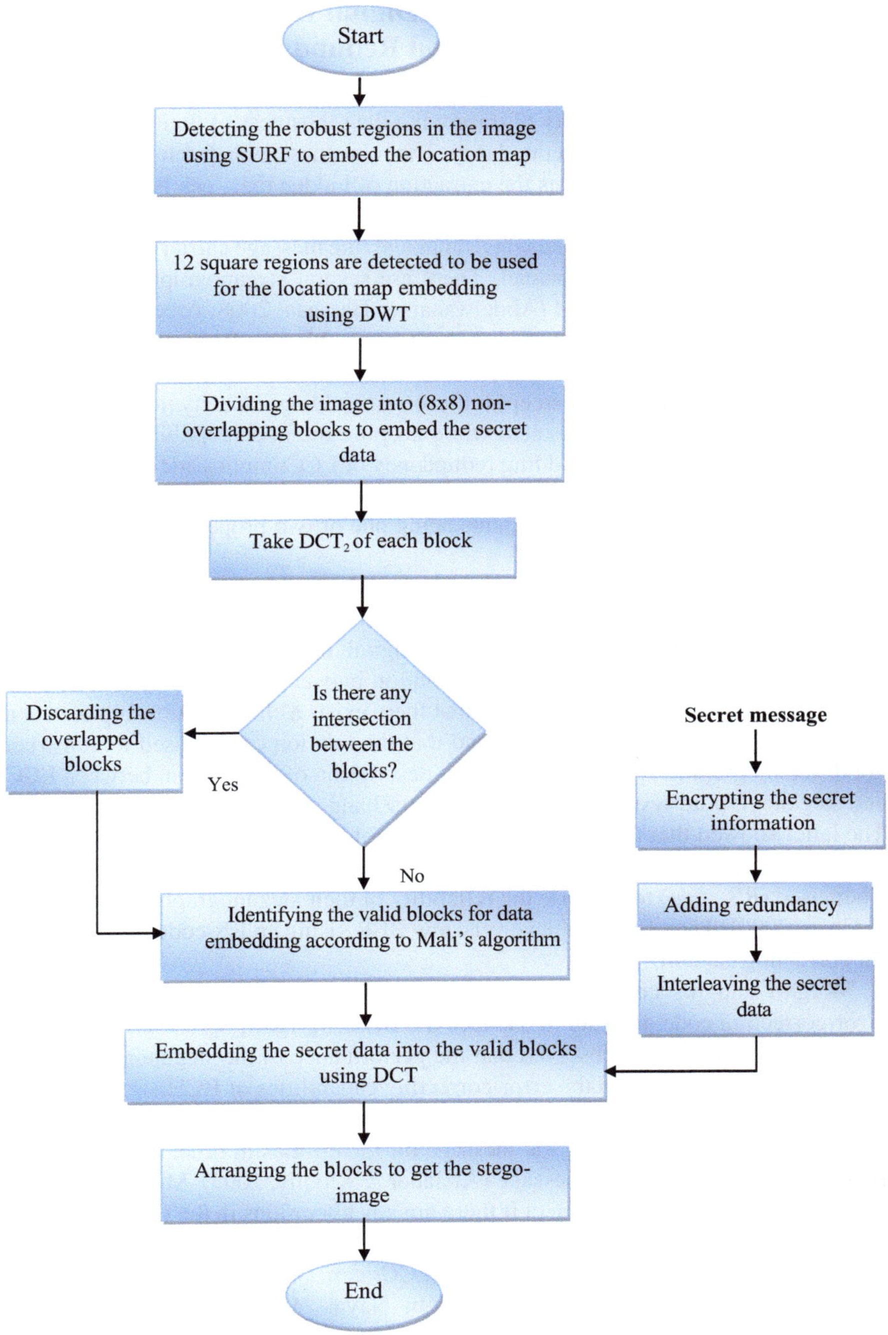

Fig. 4.5 Flow chart of the proposed IRSS algorithm

4.3 Improving the Robustness of Digital Image Steganography with ECC and Redundancy

This section discusses the ability of improving the robustness of digital image steganography by using ECC and redundancy. The robustness, which refers to how well the steganographic system resists the extraction of hidden data, can be improved by duplicating the message bits or by employing ECC techniques. Besides imperceptibility, robustness becomes crucial when an image (the message carrier) is transmitted over a public or noisy data channel. There are several steganographic techniques which can be considered robust (Abdelwahab & Hassaan, 2008; Areepongsa, Syed, Kaewkamnerd, & Rao, 2000; Hamid et al., 2012a,b; Marvel, Boncelet Jr., & Retter, 1999; Singh, Khan, Khan, & Singh, 2010; Smith & Comiskey, 1996). However, these techniques sacrifice imperceptibility or embedding capacity, and sometimes both. In order to increase robustness while keeping the imperceptibility intact, one can add redundancy or ECC. Adding redundancy or ECC implies adding some extra bits to the message, which definitely affects the actual embedding capacity. This is due to the nature of steganography; enhancing one or two properties has a negative effect on the other properties.

It is further well known that ECC introduces changes to the data to be detected and, in many cases, corrects the error. ECCs are widely used in communication and data storage to maintain the integrity of digital data (Cox, Miller, Bloom, & Honsinger, 2002). Munuera studied the relation between steganography and ECC techniques (Munuera, 2007). He found that there exists a close relationship between steganography and ECC techniques, and that the relation can be used to construct good steganographic methods. An explicit description of the relation between ECC and steganographic systems was presented by Zhang and Li in (Zhang & Li, 2005). The latter showed that there is a corresponding relationship between the Maximum Length Embeddable (MLE) codes and the perfect error-correcting codes. Lee and Chen used (RS-code) to increase the reliability of their steganographic system (Lee & Chen, 2000). They could achieve very low BER, using an RS-code with a powerful correcting capability.

Morgari and his colleagues utilized Bose-Chaudhuri-Hochquenghem (BCH) error-correcting code, in order to enhance the BER (Morgari, Spicciola, Deantonio, & Elia, 2008). They also observed that the performance of the steganographic system principally depends on the error-correcting capabilities of BCH.

Al-Jaber and Aloqily proposed their own error-checking and correcting code, which is based on duplicating the message bit three times in order to increase the probability of retrieving the message (Al-Jaber & Aloqily, 2003). Moreover, such a code enables the receiver to detect if there are any alterations in the cover image, so that it will inform the sender about these alterations. Other cases in point with respect to combining data-hiding and ECC can be found in (Giakoumaki, Pavlopoulos, & Koutsouris, 2006a, 2006b; Nayak, Bhat, Kumar, & Rajendra Acharya, 2004; Nayak, Subbanna Bhat, Acharya, & Sathish Kumar, 2009).

Adding redundancy bits is another option to increase the robustness of steganographic systems by reducing the BER. In contrast to ECC, adding redundancy bits is a simple duplication of message bits. Besides, it does not involve any mathematical equations. As a result, it can correct the errors without localizing them. The process of adding redundancy bits over ECC is characterized by its simplicity and fast computation. For the same purpose, adding redundancy bits can be used for enhancing the robustness of watermarking techniques as in (Li et al., 2011).

There are many ECC methods that can be found in the literature; however, there has been found no study that deals with the effect of ECC or adding redundancy on robustness. Accordingly, the present work is to study and compare the performances of ECC and adding redundancy when they are combined with steganography. For this purpose, RS-code, as a well-known ECC, and adding redundancy, will be combined respectively with four different data-embedding techniques. The four methods are the DWT content-based method, which is the main concern for secret data embedding in the present research work as detailed in Chap. 3, Sect. 3.3.3; DWT quantization (Deepa Kundur & Hatzinakos, 1998), DCT-based quantization (Mali et al., 2012), and histogram shifting (Zhicheng, Yun-Qing, Ansari, & Wei, 2006). The performance of each technique is measured in terms of BER after applying several types of signal-processing attacks on stego-images.

4.3.1 DWT Quantization Scheme

A DWT-based data hiding method combined with a proper quantization method was proposed by (Deepa Kundur & Hatzinakos, 1998; D. Kundur & Hatzinakos, 1999). In their scheme, the image to be used for information-hiding is transformed using Haar wavelet transform. The concept of the quantization method is based on assigning a binary number through the quantization function to every detail coefficient of the wavelet transformed image. The quantization function is given by:

$$Q(f)=\begin{Bmatrix} 0, \text{if } f/\Delta \text{ is even} \\ 1, \text{if } f/\Delta \text{ is odd} \end{Bmatrix} \tag{4.6}$$

where f is a coefficient, and Δ is the quantization parameter that is usually a positive real number. Considering the dyadic rational form of the Haar wavelet coefficients and the decreased eye sensitivity to noise in high-resolution bands, the quantization parameter Δ is defined as follows:

$$\Delta=\frac{d}{2^{c+l}} \tag{4.7}$$

where c and d are user-defined positive integers, and l is the decomposition level.

During the process of embedding, Haar wavelet decomposition of the image is performed. A secret data bit is embedded into a coefficient *f* according to the following: the quantization function is applied to the coefficient, and the resulting binary value is compared with the value of the bit to be embedded. If they are equal, the coefficient is left intact; otherwise, it is forced to cast the watermark bit value using the following assignment:

$$f = \begin{Bmatrix} f + \Delta, if\ f \leq 0 \\ f - \Delta, if\ f > 0 \end{Bmatrix} \tag{4.8}$$

Subsequently, the stego-image is produced by the corresponding level inverse wavelet transform. The extraction of the hidden information is performed through the decomposition of the stego-image, using Haar wavelet transform. Besides, the secret bits are subsequently extracted by applying the same quantization function to each of these coefficients.

4.3.2 Histogram Shifting

The method of embedding represents a lossless reversible data-hiding scheme. This is because both the message and the cover-image can be extracted exactly at the receiver's end (Fallahpour & Sedaaghi, 2007; Zhicheng et al., 2006). The algorithm is based mainly on the relocation of zeros (or minima) and peaks of the histograms of the blocks of the cover-image to embed the data; a process that leads to the modification of the gray values of some pixels. The embedding phase starts with building the histogram of the cover-image. The process of embedding an algorithm can be described as follows (Fallahpour & Sedaaghi, 2007):

Step 1: The image is divided into a series of blocks. Then steps (2–4) are iterated for each block.
Step 2: For a given number n, the same amount of pairs of peaks and zeros is recorded (P_1–P_n peaks) and (Z_1–Z_n zeros).
Step 3: The following iteration is looped n times for $i = 1 : n$.
Step 4: For each pair (P_i, Z_i):

- If $P_i > Z_i$, the image is scanned and the pixel values between $Z_i + 1$ and P_i are decremented by one. This operation shifts the histogram bins $[Z_i + 1,\ P_i]$ by one to the left, and causes a gap in the histogram bin at P_i. Then, the image is scanned again and the pixel values of $P_i - 1$ are incremented by one if the bit to be embedded is "1," or are kept intact if the bit to be embedded is "0."
- If $P_i < Z_i$, the image is scanned and the pixel values between $P_i + 1$ and $Z_i - 1$ are incremented by one. This causes a gap in the histogram bin at $P_i + 1$. Then the image is scanned again and the pixel values of P_i are incremented by one if the bit to be embedded is "1," or are kept intact if the bit to be embedded is "0." The pixel values of the zero and peak points should be transmitted along with the embedded image as side information.

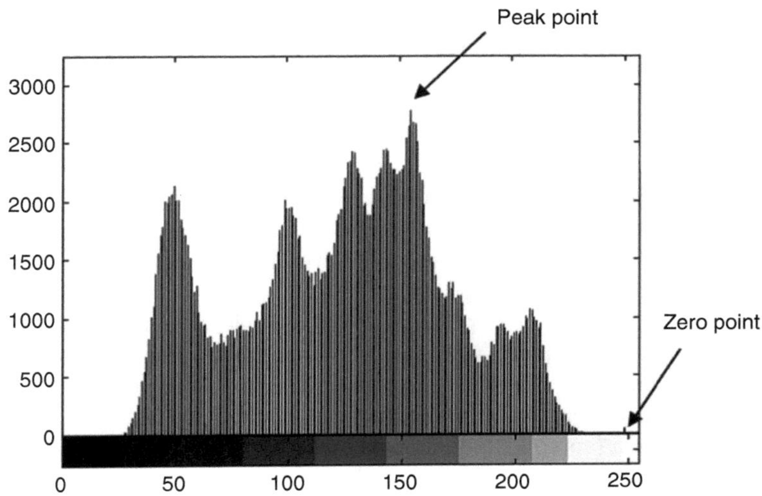

Fig. 4.6 Histogram of "Lena" image. (Zhicheng et al., 2006)

However, at the receiving end, the detection procedures are as follows:

Step 1: The stego-image is divided into a series of blocks, and the Steps (2–3) are repeated for each block.
Step 2: The following iteration is done n times for $i = 1 : n$.
Step 3: For each pair (P_i, Z_i):

- If $P_i > Z_i$, the image is scanned and the pixels with the value of P_i indicate that the embedded bit is "1." If the pixels have the value of $P_i - 1$, it indicates that the embedded bit is "0," and the pixel values should be incremented by "1." Then, all pixel values between Z_i and $P_i - 2$ are incremented by "1"; a matter which shifts the histogram to the right.
- If $P_i < Z_i$, the image is scanned and the pixels with the value of P_i indicate that the embedded bit is "0." If the pixels have the value of $P_i + 1$, it indicates that the embedded bit is "1," and the pixel values should be decremented by "1." All pixel values between $P_i - 2$ and Z_i are then decremented by "1."

It is worth mentioning that, in any case, the minima of the histogram can be used instead of zeros if there is not a sufficient number of zeros (Fallahpour & Sedaaghi, 2007). Figure 4.6 shows the histogram of the image of "Lena" (Zhicheng et al., 2006).

4.3.3 RS-Code

Reed–Solomon Code is an error-correcting code invented in 1960 by Reed and Solomon (Reed & Solomon, 1960). The first application, in 1982, of RS-codes in mass-produced products was the compact disc (CD), where two interleaved

RS-codes are used (Vries et al., 1980). Nowadays, RS-coding is very widely used in mass-storage systems to correct the burst errors associated with media defects. They are used with a hard disk drive, Digital Video Disc (DVD), telecommunication, and digital broadcast protocols (Logeshwaran & Paul, 2010).

In RS-code, a polynomial constructed from the data first undergoes oversampling. Then the polynomial is evaluated at several points, and these values are sent or recorded. Sampling the polynomial more often than is necessary makes the polynomial over-determined. As long as it receives "many" of the points correctly, the receiver can recover the original polynomial even in the presence of a "few" bad points (Reed, He, Chen, & Truong, 1998).

The principle behind error-correcting codes is simple: to transfer a sequence of symbols over a noisy channel, where the risk that nearby symbols will be transformed still occurs. The impression is to break down the sequence into blocks of a uniform length, and then one or more redundant symbols are added to each block. In the occasion of a (appropriately minor) corruption in the channel, the proposed message can be recovered by a maximum-likelihood algorithm. RS-codes are systematic linear block codes specified as RS (n, k), with m bit symbols. This means that the encoder takes k data symbols of m bits each, attaches ($n - k$) parity symbols, and produces a code word of n symbols (each of m bits) from the field Galois Field GF (2^m). The maximum block length n is equal to ($2^m - 1$) symbols for classical RS-codes (or $2m$ symbols for singly extended RS-codes). Frequently the size of the symbol is assumed to be 8 bits ($m = 8$) so that n will be 255 (Masakawa & Ochiai, 2007).

4.3.4 Adding Redundancy Bits with Interleaving

As previously discussed in Sect. 4.2.1, Mali et al. introduced a robust data-hiding system. Their system consists of secret-data processing and cover-image processing phases. In the secret-data processing phase, the message is processed by adding redundancy, combined with interleaving. This process is accomplished by making copies of the data bits to be embedded, scattered all over the stego-image (Mali et al., 2012). Therefore, although some parts of the stego-image are tampered with or changed due to noise or attacks, another copy of tampered information bits will always be available in the remaining parts of the stego-image. For a message of m bits, where each bit is repeated n times, the resultant set of bits is given by:

$$D \in \{b_{11}, b_{12}, ..b_{1n}, b_{21}, b_{22}, ..b_{2n}, \ldots b_{m1}, b_{m2}, ..b_{mn}\} \tag{4.9}$$

This is equivalent to:

$$D \in \{b_1, b_2, b_3, \ldots .b_t, b_{t+1}, b_{t+2}, b_{t+3,} \ldots, b_{2t}, \ldots, b_k\} \tag{4.10}$$

Where $k = m \bullet n$ and t is the interleaving factor.

The interleaving is achieved by arranging D on the basis of interleaving factor t as:

$$D' \in \{b_1, b_{t+1}, b_{2t+1}, \ldots, b_2, b_{t+2}, b_{2t+2}, \ldots, b_3, b_{t+3}, b_{2t+3}, \ldots, b_k\} \quad (4.11)$$

The higher the n values (the more the value of redundancy), the fewer the errors will be when recovering the information at the receiving end. However, it reduces the actual size of the data to be embedded as ECC does with a higher interleaving factor t. Consequently, more redundant bits will spread all over the image, and both redundancy and interleaving will be responsible for obtaining a robust data recovery at the receiver end.

4.4 Computer Simulation and Results

Three standard grayscale images, "Lena," "Boat," and "Gold hill," with the size of (512 × 512) are used to assess the performance of the proposed scheme explained in Sect. 4.2.2. The test images used for evaluation are shown in Fig. 4.7.

The proposed algorithm, IRSS, is compared to Mali et al.'s algorithm in order to evaluate the reliability with different levels of certain attacks applied to the stego-images, such as JPEG compression and AWGN. Tables 4.2 and 4.3 introduce a comparison between the IRSS algorithm and Mali et al.'s algorithm in terms of robustness versus JPEG compression and AWGN attacks. The embedding methodology in the IRSS scheme is tuned to JPEG compression. Thus, the decoding process of the embedded data is perfect for all JPEG compression attacks that are less than or equal to the given QF. Likewise, the method considered in this work is resilient to AWGN attack with (100%) recovery for the embedded information.

'Lena' 'Boat' 'Gold hill'

Fig. 4.7 The standard images used for evaluation

Table 4.2 A comparison between the IRSS algorithm and Mali's algorithm in terms of reliability (with QF = 75% and $w^{\wedge} = 0.5$)

	Number of misidentified blocks					
	IRSS algorithm			Mali's algorithm		
Type of attack	Lena	Boat	Gold hill	Lena	Boat	Gold hill
No attack	0	0	0	27	15	17
JPEG 100%	0	0	0	−14	−9	−5
JPEG 80%	0	0	0	316	249	228
AWGN (45 dB)	0	0	0	−26	−12	−5
AWGN (35 dB)	0	0	0	−32	−12	2

Table 4.3 A comparison between the IRSS algorithm and Mali's algorithm in terms of reliability (with QF = 75% and $w^{\wedge}= 0.8$)

	Number of misidentified blocks					
	IRSS algorithm			Mali's algorithm		
Type of attack	Lena	Boat	Gold hill	Lena	Boat	Gold hill
No attack	0	0	0	23	14	32
JPEG 100%	0	0	0	−12	−3	6
JPEG 80%	0	0	0	250	247	175
AWGN (45 dB)	0	0	0	−31	−12	10
AWGN (35 dB)	0	0	0	−28	−20	2

It is important to note that the number of the overlooked blocks can be considered by deducting the number of the detected blocks in the receiving end from the blocks that have been used for secret data embedding (the actual embedding blocks). Consequently, when the number of the detected blocks is less than the actual number of the embedding blocks, the outcome will be a positive number. In contrast, as a result of the functional attacks to the stego-image, erroneous blocks may be detected. Therefore, the number of detected blocks is more than the actual number of the embedding blocks. This is the reason that some numbers have negative values in Tables 4.2 and 4.3.

In terms of imperceptibility, the visual quality of the obtained stego-image with the IRSS algorithm is better compared to those obtained by Mali et al.'s algorithm as illustrated in Table 4.4. The visual quality is measured by the PSNR, as given in Eq. (2.6) in (Cover, 2006). The obtained results demonstrate that the IRSS algorithm has a better visual quality. For the purpose of evaluation, another comparison is introduced in terms of the hiding capacity. This capacity is measured as the number of payload bits that can be embedded in the image and retrieved successfully (with 100% recovery). The obtained results are shown in Table 4.5.

As discussed in Sect. 4.3, to study the effect of adding redundancy to the secret message compared with inserting ECC on enhancing the stego-system robustness, four algorithms, DWT content-based method, DWT quantization, DCT-based quantization, and histogram shifting technique, are implemented and applied to six

Table 4.4 A comparison between the IRSS algorithm and Mali's algorithm in terms of visual quality (PSNR)

	PSNR values (dB) with QF = 75%					
	IRSS algorithm			Mali's algorithm		
Energy thresholding w^=	Lena	Boat	Gold hill	Lena	Boat	Gold hill
0.5	37.28	34.10	36.42	33.19	32.66	33.36
0.6	37.62	34.24	36.98	33.56	32.78	33.82
0.7	38.33	34.67	37.61	33.91	32.94	34.28
0.8	38.37	35.10	38.51	34.22	33.18	34.89
0.9	38.79	35.74	39.18	34.69	33.53	35.58
1	39.74	36.76	39.53	35.31	34.16	36.57

Table 4.5 A comparison between the IRSS algorithm and Mali's algorithm in terms of hiding capacity

	Hiding capacity (bits)					
	IRSS algorithm			Mali's algorithm		
Energy thresholding w^=	Lena	Boat	Gold hill	Lena	Boat	Gold hill
0.5	25,536	46,144	30,016	83,664	93,996	80,304
0.6	22,400	44,352	25,536	75,936	91,196	72,072
0.7	19,264	38,976	21,056	70,476	87,892	64,344
0.8	18,816	33,152	16,128	65,688	83,272	56,196
0.9	15,680	25,088	12,096	58,884	76,580	47,796
1	11,648	16,128	10,752	51,212	66,724	38,444

standard grayscale images with size 512 × 512 pixels, as shown in Fig. 4.8. The data that has been used is a randomly generated message of 1000 bits. The RS-codes that have been used are those which produce correction bits that are equal to the bit produced by a specific redundancy factor n as illustrated in Table 4.6. Those codes have been selected with some approximation, because sometimes it is hard to get the RS-code that gives the exact amount that is equivalent to the number of redundancy bits. Due to the limitation of the hiding capacity of the histogram shifting method discussed in Sect. 4.3.2, only redundancy factors (values) whose values are up to 8 have been used. The interleaving factor has been set to (50) throughout all the tests. The robustness of the embedding methods has been measured in terms of BER. The average of BER (for the 6 images) is given in Tables 4.7, 4.8, 4.9, 4.10, and 4.11 and is illustrated in Figs. 4.9, 4.10, 4.11, and 4.12.

4.5 Discussion and Analysis

The algorithm recommended by (Mali et al., 2012) displays satisfactory levels of robustness as a result of merging DCT and addition of redundancy bits. Conversely, some of the blocks that carry data are misidentified throughout the removal course.

Fig. 4.8 Standard images used for testing

Table 4.6 The RS-codes and their equivalent *n* values

Redundancy factor (*n*)	Equivalent RS-Code
2	(15,5)
4	(15,3)
6	(63,9)
8	(63,7)
10	(127,11)
12	(255,19)
14	(255,17)

Table 4.7 The percentage BER without using ECC or redundancy

Type of attack	DWT-based quantization	DWT content-based	DCT-based quantization	Histogram shifting
JPEG compression (QF = 100%)	5.41	18.60	5.5825	48.73
JPEG compression (QF = 80%)	40.30	33.11	23.76333	48.73
AWGN (35 dB)	42.12	30.50	17.23833	50.06
AWGN (25 dB)	49.92	36.92	34.72667	49.95
Median filter (3 × 3)	42.72	34.44	47.54167	49.54
Low pass filter (3 × 3)	42.04	35.88	45.91417	49.40

Table 4.8 (**a**) Percentage BER versus RS-code with DWT-based quantization method (**b**) Percentage BER versus redundancy with DWT-based quantization method

(a) RS-codes							
Type of attack	(15,5)	(15,3)	(63,9)	(63,7)	(127,11)	(255,19)	(255,17)
JPEG compression (QF = 100%)	0.70	0.24	0.05	0.03	0.00	0.00	0.01
JPEG compression (QF = 80%)	41.10	41.63	41.79	42.77	41.96	41.85	42.49
AWGN (35 dB)	42.04	41.96	42.03	41.90	41.72	41.80	42.12
AWGN (25 dB)	49.86	50.12	49.91	49.78	49.68	50.11	50.13
Median filter (3 × 3)	43.74	44.41	44.84	45.17	44.69	44.69	44.70
Low pass filter (3 × 3)	43.32	44.36	44.65	44.83	45.18	44.26	44.27
(b) Redundancy factor (n)							
Type of attack	2	4	6	8	10	12	14
JPEG compression (QF = 100%)	0.99	0.17	0.04	0.00	0.00	0.00	0.00
JPEG compression (QF = 80%)	36.82	35.03	33.05	31.82	30.82	29.45	28.44
AWGN (35 dB)	17.60	16.55	15.91	15.41	14.73	14.22	9.50
AWGN (25 dB)	49.83	50.41	49.81	50.09	49.98	49.92	50.39
Median filter (3 × 3)	41.23	39.66	39.78	39.32	38.51	38.23	37.58
Low pass filter (3 × 3)	41.74	41.04	41.00	40.37	39.97	38.50	38.07

Table 4.9 (**a**) Percentage BER versus RS-code with DWT content-based method (**b**) Percentage BER versus redundancy with DWT content-based method

(a) RS-codes							
Type of attack	(15,5)	(15,3)	(63,9)	(63,7)	(127,11)	(255,19)	(255,17)
JPEG compression (QF = 100%)	0.00	0.00	0.00	0.00	0.00	0.00	0.00
JPEG compression (QF = 80%)	10.20	7.33	12.50	11.91	12.82	13.23	12.83
AWGN (35 dB)	4.24	2.37	4.25	3.12	4.85	7.29	6.65
AWGN (25 dB)	27.86	26.96	27.30	26.34	26.81	26.94	26.49
Median filter (3x3)	18.32	17.99	20.39	22.03	21.47	22.12	22.96
Low pass filter (3x3)	14.97	14.06	16.98	18.15	17.73	18.56	19.64
(b) Redundancy factor (n)							
Type of attack	2	4	6	8	10	12	14
JPEG compression (QF = 100%)	8.52	4.35	2.41	1.41	0.90	0.61	0.39
JPEG compression (QF = 80%)	25.71	20.67	16.78	13.87	11.22	9.58	8.14
AWGN (35 dB)	21.50	16.24	12.39	9.59	7.45	5.82	4.73
AWGN (25 dB)	30.55	26.16	22.20	19.10	16.24	14.04	12.67
Median filter (3x3)	32.40	29.28	26.93	25.02	22.53	21.08	20.01
Low pass filter (3x3)	32.61	29.81	27.46	25.03	22.86	21.23	19.34

Table 4.10 (**a**) Percentage BER versus RS-code with DCT-based quantization method (**b**) Percentage BER versus redundancy with DCT-based quantization method

(a) RS-codes							
Type of attack	(15,5)	(15,3)	(63,9)	(63,7)	(127,11)	(255,19)	(255,17)
JPEG compression (QF = 100%)	1.84	1.18	0.01	0.01	0.00	0.00	0.00
JPEG compression (QF = 80%)	22.59	21.11	21.12	20.55	20.18	19.24	19.88
AWGN (35 dB)	15.32	12.94	16.26	17.99	15.14	15.12	15.65
AWGN (25 dB)	35.36	35.16	33.81	36.58	32.17	31.86	33.12
Median filter (3x3)	47.69	47.91	46.42	45.59	43.73	43.53	45.45
Low pass filter (3x3)	46.35	46.24	44.60	43.52	41.90	41.27	43.17
(b) Redundancy factor (*n*)							
Type of attack	2	4	6	8	10	12	14
JPEG compression (QF = 100%)	0.78	0.13	0.05	0.01	0.00	0.00	0.00
JPEG compression (QF = 80%)	14.03	8.98	6.84	4.88	3.04	2.58	1.67
AWGN (35 dB)	7.54	3.66	2.75	1.06	0.51	0.35	0.17
AWGN (25 dB)	28.33	23.81	23.29	19.26	16.87	16.34	14.53
Median filter (3x3)	46.86	46.82	46.55	46.82	46.80	47.02	47.34
Low pass filter (3x3)	44.46	43.98	43.37	43.20	42.70	42.70	43.34

Table 4.11 (**a**) Percentage BER versus RS-code with histogram shifting method (**b**) Percentage BER versus redundancy with histogram shifting method

(a) RS-codes				
Type of attack	(15,5)	(15,3)	(63,9)	(63,7)
JPEG compression (QF = 100%)	49.26	49.32	49.20	49.42
JPEG compression (QF = 80%)	50.02	50.03	50.01	49.93
AWGN (35 dB)	49.79	50.24	49.87	49.90
AWGN (25 dB)	50.11	50.14	49.92	50.17
Median filter (3x3)	49.84	49.75	50.06	49.98
Low pass filter (3x3)	50.44	50.15	49.93	49.93
(b) Redundancy factor (*n*)				
Type of attack	2	4	6	8
JPEG compression (QF = 100%)	47.36	46.88	46.97	46.86
JPEG compression (QF = 80%)	47.31	46.88	46.88	46.88
AWGN (35 dB)	49.96	49.95	49.94	49.94
AWGN (25 dB)	49.95	50.08	49.88	49.98
Median filter (3x3)	49.38	49.27	49.17	49.26
Low pass filter (3x3)	49.96	50.14	50.02	49.83

The current work targets improving the consistency of the original algorithm by overcoming the problem of the misidentified blocks (Hamid et al., 2012a, 2013b). To do so, an embedding map (location map) has been adopted to indicate the location of the blocks that have been used for embedding. This means that some regions

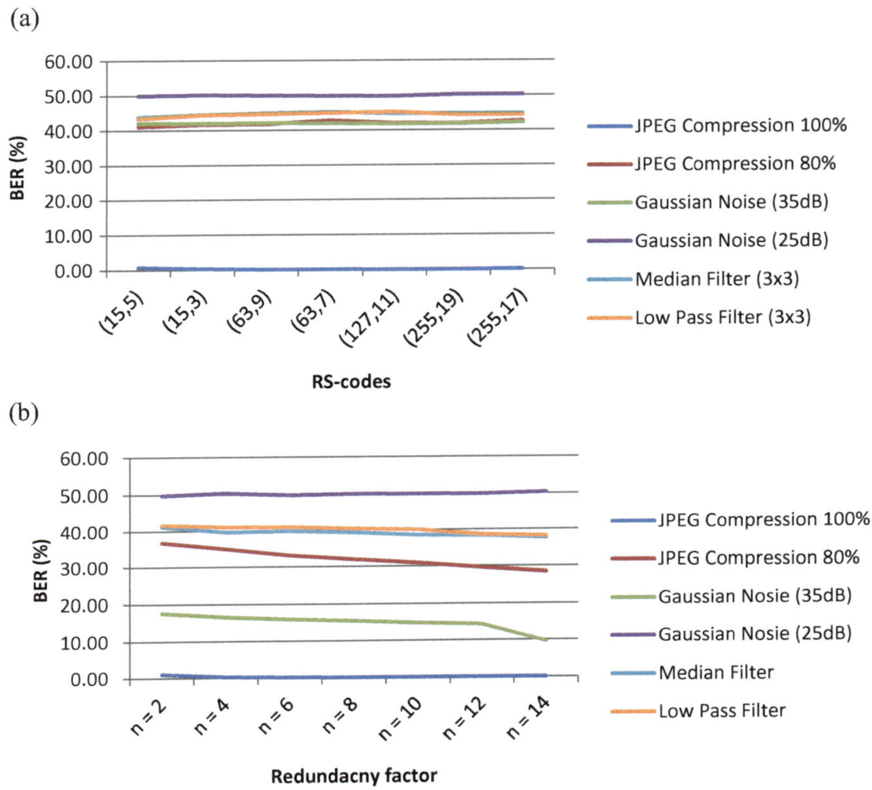

Fig. 4.9 BER for DWT quantization method (**a**) with RS-code (**b**) with adding redundancy

of the image will be exploited to hide data, while others will be used to hide the embedding map, as shown in Fig. 4.4 and clarified by the flow chart given in Fig. 4.5. The blocks in which the data are concealed are determined according to Mali et al.'s algorithm. On the other hand, the regions in which the embedding map is concealed are determined in an extremely dynamic way to increase the security of the algorithm. This goal has been achieved by using the SURF technique to find the robust interest points of the image. This is because each image has different interest points. The embedding map is incorporated in those regions using the DWT content-based method. The experimental results given in Tables 4.2 and 4.3 show that the present proposed algorithm, namely IRSS, can overcome the problem of block misidentification, even when the stego-image undergoes JPEG compression or Gaussian noise. Such a solution has been developed by inserting the embedding map, and by choosing the most robust regions of the image for embedding this map frequently. The high reliability of the IRSS algorithm in identifying the blocks comes from the ability of SURF to detect the interest points, even after applying the attacks. Besides, the DWT-based embedding technique also plays an important role in keeping the embedding map intact.

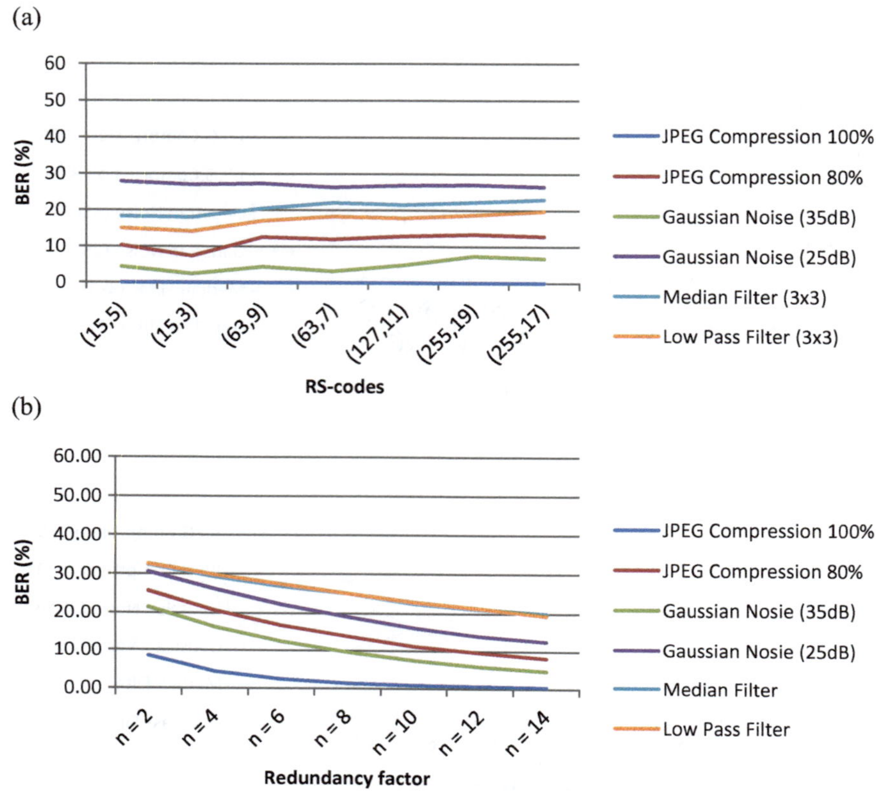

Fig. 4.10 BER for DWT with content-based method (**a**) with RS-code (**b**) with adding redundancy

In comparison to Mali et al.'s algorithm, the IRSS algorithm shows a better visual quality in terms of PSNR, as shown in Table 4.4. This can be said for the three images used for testing and with all ($w^{\wedge}$) values used. In Mali et al.'s algorithm, the embedded message bits are extracted sequentially from the blocks without any prediction of the blocks that contain the message bits. Due to the nature of the extraction process in Mali et al.'s algorithm, all the available capacity should be used in the process of embedding. Even when the message size is less than the available capacity, zeros are padded to the message to make its size equal to the available capacity. On the other hand, the IRSS algorithm exploits an embedding map that determines the exact locations of the blocks that contain the message bits. Consequently, only the necessary capacity is used for embedding the message, not all of it. This helps in reducing the overall size of the message and hence enhances the imperceptibility of the stego-image. In addition, from the obtained results in Table 4.4, one can deduce that the value of PSNR increases with the increase of $w^{\wedge}$.

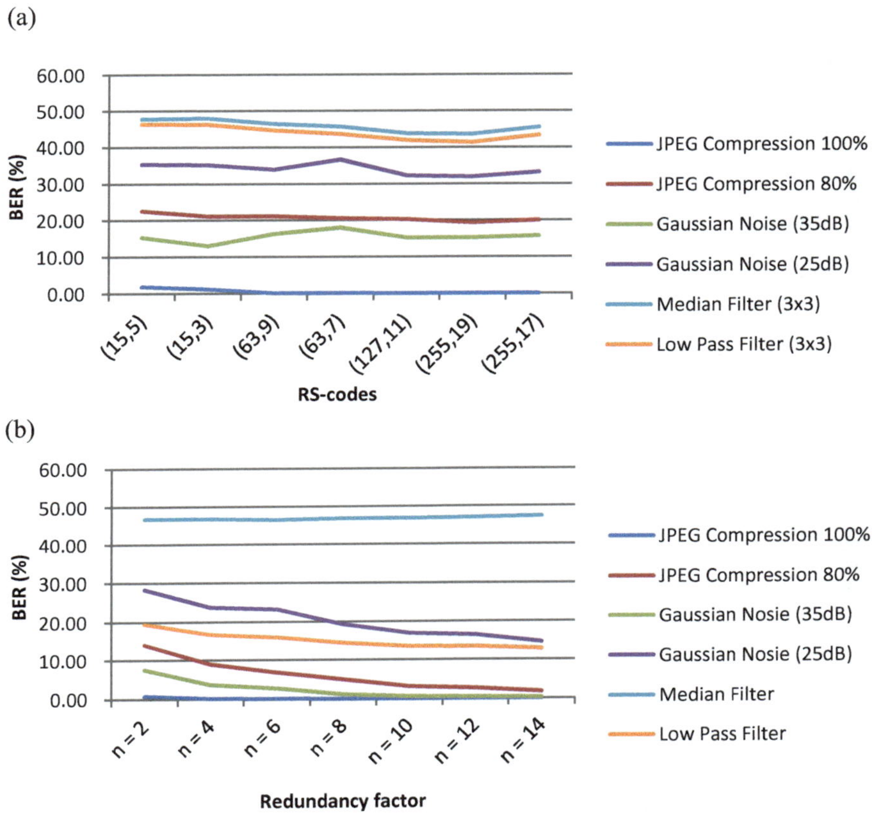

Fig. 4.11 BER for DCT-quantization method (**a**) with RS-code (**b**) with adding redundancy

This is because of the trade-off between the image quality and the volume of embedding at a given robustness (determined by the selected QF).

However, Table 4.5 indicates that the hiding capacity of the IRSS scheme is somewhat lower than that achieved by Mali et al.'s algorithm. The capacity reduction can be attributed to the exploitation of some regions of the image for the purpose of hiding the embedding map. The robustness of the IRSS algorithm in terms of BER is not tackled in the present work. This is because the same embedding technique used in Mali et al.'s algorithm for hiding the data has been used in the present study; this means that both algorithms have the same level of robustness.

As a matter of fact, the robustness of a steganographic system is a crucial factor; especially when an unsecured or noisy channel is used for transmission. Accordingly, this chapter introduces a study on the possibility of improving stego-system robustness in terms of BER reduction. Such a step can be achieved either by inserting ECC or by adding redundancy. For the DWT-based quantization method, the obtained experimental results show that adding redundancy ($n = 14$) can significantly enhance robustness; especially with 100% JPEG compression or low levels of Gaussian

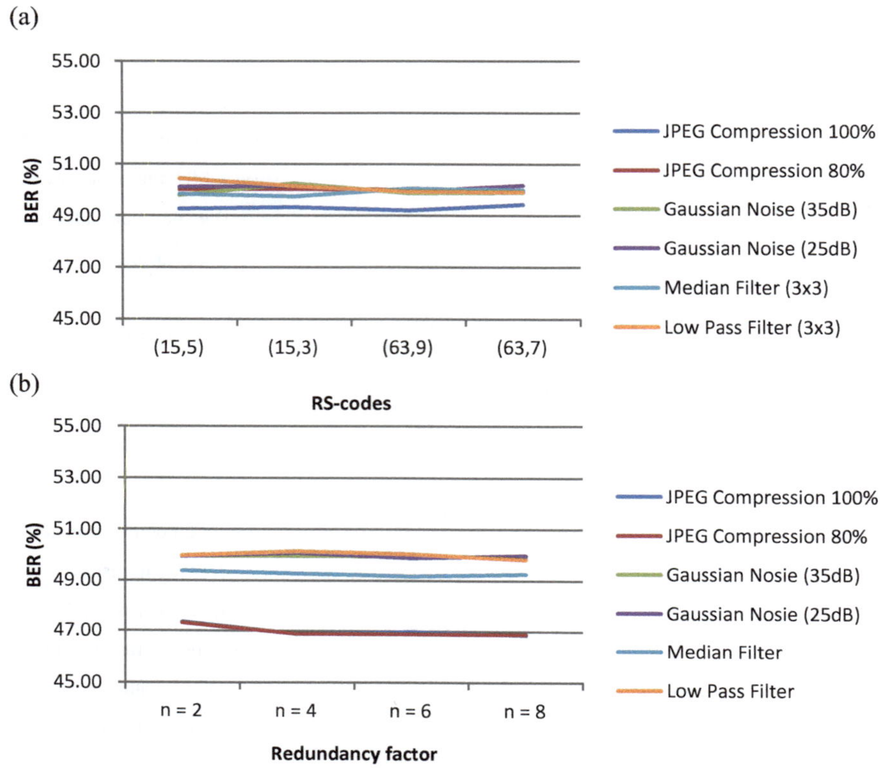

Fig. 4.12 BER for histogram shifting method (**a**) with RS-code (**b**) with adding redundancy

noise, as shown in Table 4.8b. It can enhance the robustness with median or low-pass filters, as is the case with the BER values that have been reduced significantly. However, the BER values are still high (37.58%, 38.07%). Referring to Table 4.8a, it has been noticed that the RS-codes which have been used in the comparison did not exhibit the same performance of adding redundancy except when 100% JPEG compression was applied. In addition, it was also noticed that adding RS-codes decreased robustness. This was due to the fact that the BER values were higher than the corresponding values when neither ECC nor redundancy was used, as shown in Fig. 4.9.

The same conclusion applies to the DCT quantization method. That is, adding redundancy can considerably improve the robustness of the stego-system; this has in return been inferred by the reduction of the BER values, as indicated by the values given in Table 4.10b and illustrated in Fig. 4.11. However, low-pass and median filters still represent unbeatable attacks for both RS-codes and adding redundancy, as denoted by the BER values of (43.17%, 47.34%).

In contrast, the experimental results showed that DWT with content-based manner worked well with RS-code (15, 3). It has better performance than adding redundancy even with $n = 12$, as shown in Tables 4.9a and b and Fig. 4.10. The RS-code (15, 3) also reduces the BER values when low-pass and median filters are applied, up to 14.06% and 17.99% respectively.

The histogram shifting method is not robust, as is indicated by the high BER values given in Table 4.7. Neither exploiting RS-codes nor adding redundancy could enhance robustness by even a little, as shown in Tables 4.11a and b, respectively, and demonstrated in Fig. 4.12. However, the simulated results indicate that RS-codes and the addition of redundancy can enhance the robustness of the embedding methods for limited levels. The level of enhancement depends on the RS-code, the redundancy factor which has been selected, and on the native robustness of the embedding methods. In other words, one cannot say that RS-code is better than adding redundancy or vice versa. This is because for certain types of embedding methods, RS-code may give better performance, and for other methods adding redundancy is considered the best. However, for some embedding methods that do not have minimum levels of robustness, robustness could be enhanced by adding RS-codes or redundancy.

4.6 Summary

This chapter deals with two main aspects, the first of which is concerned with introducing an improved robust and secure image steganography system. The original algorithm presented by Mali et al. gives adequate results in terms of robustness; however, the algorithm cannot be considered reliable, as some data is lost. In order to overcome the problem of the lost data due to the misidentified blocks, an embedding map was proposed to specify the location of the blocks that can be used for embedding a secret message. Such a technique, the embedding map, is very important if one needs to start the extracting phase correctly and accurately. Besides, any loss in the embedding map will in turn lead to data loss. Consequently, the embedding map will be hidden using SURF and DWT to assure robustness and to increase the level of security of the proposed system. The secret data can be embedded using the original algorithm (DCT-based) proposed by Mali et al. More so, the experimental results show the ability of the new algorithm, namely IRSS, to overcome the problem of losing data even with JPEG compression or Gaussian noise.

The second focus of this chapter aims to study the effect of inserting the RS-code, a well-known ECC, on the robustness of an image steganography system. Then the process of inserting will be compared with that of adding redundancy to the secret data to be embedded in the cover image. For this purpose, RS-code and adding redundancy are combined respectively with four different data-embedding techniques. These methods include DCT quantization, DWT quantization, DWT content-based, and histogram shifting. In terms of BER, the experimental results show that adding redundancy bits to the message has a much better effect on improving the robustness of the first two embedding methods. In contrast, the RS-code (15, 3) improves the third embedding method more than the process of adding redundancy. However, neither the RS-code nor the addition of redundancy could improve the fourth embedding method. As a result, it is obvious that not all steganographic systems can be enhanced by employing ECC. However, adding redundancy may be a good choice for some steganographic systems.

References

Abdul-mahdi, N. H., Yahya, A., Ahmad, R. B., & Al-Qershi, O. M. (2013). Secured and robust information hiding scheme. *Procedia Engineering Journal, 53*, 463–471.

Abdelwahab, A. A., & Hassaan, L. A. (2008, March 18–20). *A discrete wavelet transform based technique for image data hiding*. Paper presented at the radio science conference, 2008. NRSC 2008. National.

Al-Jaber, A., & Aloqily, I. (2003). High quality steganography model with attacks detection. *Information Technology Journal, 2*, 116–127.

Areepongsa, S., Syed, Y. F., Kaewkamnerd, N., & Rao, K. R. (2000). *Steganography for a low bit-rate wavelet based image coder.* Paper presented at the image processing, 2000. Proceedings. 2000 International Conference on.

Alattar, A. M. (2004). Reversible watermark using the difference expansion of a generalized integer transform. *IEEE Transactions on Image Processing, 13*(8), 1147–1156.

Cover, K. S. (2006). *Multiexponential reconstruction algorithm immune to false positive peak detection.* Review of Scientific Instruments 77,075101 (2006); https://doi.org/10.1063/1.2206780.

Cox, I., Miller, M., Bloom, J., & Honsinger, C. (2002). Digital Watermarking. *Journal of Electronic Imaging, 11*(3), 414–414.

Fallahpour, M., & Sedaaghi, M. H. (2007). High capacity lossless data hiding based on histogram modification. *IEICE Electronics Express, 4*(7), 205–210.

Giakoumaki, A., Pavlopoulos, S., & Koutsouris, D. (2006a). Multiple image watermarking applied to health information management. *IEEE Transactions on Information Technology in Biomedicine, 10*(4), 722–732.

Giakoumaki, A., Pavlopoulos, S., & Koutsouris, D. (2006b). Secure and efficient health data management through multiple watermarking on medical images. *Medical and Biological Engineering and Computing, 44*(8), 619–631.

Hamid, N., Yahya, A., Ahmad, R. B., & Al-Qershi, O. M. (2012a). Image steganography techniques: An overview. *International Journal of Computer Science and Security (IJCSS), 6*(3), 168–178.

Hamid, N., Yahya, A., Ahmad, R. B., & Al-Qershi, O. M. (2012b, April). *Characteristic region based image steganography using speeded-up robust features technique*. Paper presented at the 1st International Conference on Future Communication Network (ICFCN'12). IEEE international conference, Iraq, Baghdad.

Hamid, N., Yahya, A., Ahmad, R. B., Najim, D., & Kanaan, L. (2013a). Steganography in image files: A survey. *Australian Journal of Basic and Applied Sciences, 7*(1), 35–55.

Hamid, N., Yahya, A., Ahmad, R. B., Najim, D., & Kanaan, L. (2013b). Enhancing the robustness of digital image steganography using ECC and redundancy. *WULFENIA Journal, 20*(4), 153–169.

Jun, T. (2003). Reversible data embedding using a difference expansion. *IEEE Transactions on Circuits and Systems for Video Technology, 13*(8), 890–896.

Kundur, D., & Hatzinakos, D. (1998). *Towards a telltale watermarking technique for tamper-proofing*. Image Processing, 1998. ICIP 98. Proceedings. Chicago, IL, USA, USA.

Kundur, D., & Hatzinakos, D. (1999). Digital watermarking for telltale tamper proofing and authentication. *Proceedings of the IEEE, 87*(7), 1167–1180.

Lee, Y-K., & Chen, L-H. (2000). *A secure robust image steganographic model*. Paper presented at the tenth national conference of information security.

Li, L., Qian, J., & Pan, J. S. (2011). Characteristic region based watermark embedding with RST invariance and high capacity. *AEU - International Journal of Electronics and Communications, 65*(5), 435–442.

Logeshwaran, R., & Paul, I. J. L. (2010, January). *Performance study on the suitability of reed solomon codes in WiMAX*. Paper presented at the wireless communication and sensor computing, 2010. International Conference on ICWCSC 2010.

Mali, S. N., Patil, P. M., & Jalnekar, R. M. (2012). Robust and secured image-adaptive data hiding. *Digital Signal Processing: A Review Journal, 22*(2), 314–323.

Malik, F., & Baharudin, B. (2013). The statistical quantized histogram texture features analysis for image retrieval based on median and laplacian filters in the DCT domain. *International Arab Journal of Information Technology, 10*(6), 1–9.

Marvel, L. M., Boncelet Jr., C. G., & Retter, C. T. (1999). Spread spectrum image steganography. *IEEE Transactions on Image Processing, 8*(8), 1075–1083.

Masakawa, T., & Ochiai, H. (2007, March). *Design of reed-solomon codes for OFDM systems with clipping and filtering*. Paper presented at the wireless communications and networking conference, 2007. WCNC 2007. IEEE.

Morgari, G., Spicciola, M., Deantonio, S., & Elia, M. (2008). Steganographic schemes for noisy communication channels. *Journal of Contemporary Engineering Sciences, 1*(1), 27–39.

Munuera, C. (2007). Steganography and error-correcting codes. *Signal Processing, 87*(6), 1528–1533.

Nayak, J., Bhat, P. S., Kumar, M. S., & Rajendra Acharya, U. (2004). *Reliable transmission and storage of medical images with patient information using error control codes*. India Annual Conference, 2004. Proceedings of the IEEE INDICON 2004, Kharagpur, India.

Nayak, J., Subbanna Bhat, P., Acharya, U. R., & Sathish Kumar, M. (2009). Efficient storage and transmission of digital fundus images with patient information using reversible watermarking technique and error control codes. *Journal of Medical Systems, 33*(3), 163–171.

Reed, I. S., He, R., Chen, X., & Truong, T. K. (1998). Application of Grobner bases for decoding Reed-Solomon codes used on CDs. *IEEE Proceedings – Computers and Digital Techniques, 145*(6), 369–376.

Reed, I. S., & Solomon, G. (1960). Polynomial codes over certain finite fields. *Journal of the Society for Industrial and Applied Mathematics, 8*(2), 300–304.

Singh, R. P., Khan, M. A. A., Khan, M., & Singh, N. (2010). Spread spectrum image steganography in multimedia messaging service of mobile phones. *International Journal of Electronics Engineering, 2*(2), 365–369.

Smith, J. R., & Comiskey, B. O. (1996). *Modulation and information hiding in images*. Paper presented at the proceedings of the first international workshop on information hiding.

Vries, L. B., Immink, K. A., Nijboer, J. G., Hoeve, H., Philips, N. V., Doi, T. T., & Odaka, K. (1980). *The compact disc digital audio system: Modulation and error correction*. Paper presented at the 67th AES Convention, No. 1674(H-8).

Zhang, W., & Li, S. (2005). Steganographic codes — a new problem of coding theory. *Journal of Latex Class Files, 1*(11), 7.

Zhicheng, N., Yun-Qing, S., Ansari, N., & Wei, S. (2006). Reversible data hiding. *IEEE Transactions on Circuits and Systems for Video Technology, 16*(3), 354–362.

Chapter 5
Conclusion Toward Hidden Communication

Abstract Steganography is the art and science of invisible communication. It has recently been the core of attention by the information-security community. This chapter is concerned with summarizing the conclusions that have been arrived at throughout conducting the present work. It further showcases some possible future research challenges that need to be studied in this area.

5.1 Conclusion

Steganography is the art and science of invisible communication. It has recently been the core of attention by the information-security community. This chapter is concerned with summarizing the conclusions that have been arrived at throughout conducting the present work (Hamid et al., 2012a,b, 2013a,b). It further showcases some possible future research challenges that need to be studied in this area.

As is clearly indicated in Chap. 1, the main objective of this book is to investigate a novel approach to image steganography that helps provide enhancements to the existing steganography algorithms. Accordingly, the focus of the present research was not only on the embedding strategy, as is the case in most of the research studies, but also on the pre-processing stages, such as payload encryption and embedding-region selection. The researcher, throughout the whole work, has focused on the following main points:

First, a comprehensive review of previous work in digital-image steganography was discussed and classified into four main categories based on the cover modifications caused by the embedding method. In addition, advantages and disadvantages of algorithms within each category have been highlighted and discussed where possible. Throughout examining the previous studies, the researcher has recognized that almost all the current algorithms rely heavily on exploiting all the cover-image pixels to hide the secret information. Moreover, during the process of searching for non-smooth regions for embedding, it was noticed that the introduced algorithms were time-consuming and inefficient. These two facts drove the researcher toward

A. Yahya, *Steganography Techniques for Digital Images*,
https://doi.org/10.1007/978-3-319-78597-4_5

achieving some contributions, such as the ones reported in this book (Hamid et al., 2012a,b, 2013a,b).

Second, a new algorithm named CR-BIS was investigated. Such an algorithm uses the characteristic-regions embedding strategy for hiding secret information in digital images. Two characteristics, imperceptibility and robustness, were employed to evaluate the performance of the proposed techniques. Moreover, error-free recovery of the embedded secret data was enforced without referring to the original cover-image. The results were promising and outperformed any relevant methods. CR-BIS embeds data in the wavelet domain. This technique has proven to be robust and causes less distortion to the carrier file. Security tests applied to verify the strength of the algorithm included JPEG compression, AWGN, image filtering, and salt and pepper noise. This book adopts the technique of avoiding smooth regions to embed the secret bits using SURF. The latter scans the cover-image to detect the invariant robust regions to be used for embedding secret data. The embedding process usually takes place in the wavelet domain using the content-based method and is guided by the robust regions decided by the SURF technique. Accordingly, it was noticed that the process of hiding information using a content-based manner adds another level of security and robustness. This is because this algorithm tends to modify the horizontal and vertical wavelet coefficients of the cover-image. It creates a balance between the high frequencies, which can be removed by even minor JPEG compression, and low frequencies in which the modification is easily seen by the human eye. A summary of the drawbacks of the current steganographic techniques in the literature and a list of the main characteristics underlying the proposed method of this thesis are summarized in Appendix A.

Third, as far as the objective of building up a robust yet reliable steganographic system is concerned, another contribution has been conceived. The developed stego-system was introduced by exploiting the benefits brought by the proposed CR-BIS algorithm. However, it was noticed that the stego-system proposed by (Mali, Patil, & Jalnekar, 2012) cannot be considered as a reliable and applicable steganographic system. This is because although this system can successfully embed the secret information into the DCT coefficients, it cannot retrieve the hidden information with 100% recovery. Consequently, the researcher of the present thesis has developed a new scheme called IRSS, which involves inserting an embedding map to indicate the correct locations of the embedded secret information. The embedding map was frequently embedded in the most robust regions of the image indicated by SURF. Such a process was done by modifying the wavelet coefficients of these robust regions in a content-based manner. The simulated results have shown that by using the IRSS technique, the embedded secret data can be extracted completely from the resulting stego-image without any error. Moreover, a better visual quality can be obtained for the resultant stego-images, as is clearly indicated by the higher obtained values of the PSNR.

Fourth, the capability of enhancing the stego-system robustness by either inserting ECC or adding redundancy to the embedded message has also been considered. In addition to the DWT content-based method, which is the main concern of the current research work, three well-known embedding techniques have been implemented. These are DWT quantization (Deepa Kundur & Hatzinakos, 1998), DCT-based quantization (Mali et al., 2012), and histogram shifting (Zhicheng, Yun-Qing, Ansari, & Wei, 2006). As a popular ECC, the RS-Code (Irving S. Reed & Solomon, 1960) was used to produce correction bits that are equal to the number of bits produced by a specific redundancy factor *n*.

However, through the experimental simulation, it has been found that adding redundancy bits to the message has a much better effect on improving the robustness of both the DCT quantization and DWT quantization schemes. In contrast, the RS-code (15,3) improved the DWT content-based embedding method more than the addition of redundancy. But neither the RS-code nor the addition of redundancy could improve the histogram-shifting embedding technique.

5.2 Steganographic Contributions

In this book, two different steganographic algorithms have been proposed and the contribution of this book can be summarized as follows (Hamid et al., 2012a,b, 2013a,b):

- The first algorithm, CR-BIS, combined the robustness of SURF and DWT in order to achieve characteristic region steganography synchronization. In addition to the feature of robustness that the developed algorithm is characterized with, it further helped avoid hiding data in the whole image by selecting characteristic regions for embedding dynamically. Such a dynamic manner of selecting regions for embedding has increased the security and the imperceptibility as well; this was indicated by the obtained high PSNR values, up to 48.30.
- The second algorithm, IRSS, is an improvement of the existing algorithm proposed by Mali et al. The latter algorithm had a reliability defect, as some data could not be retrieved at the extraction phase. The IRSS algorithm has overcome this problem by adopting the concept of the embedding map that refers to the regions used for embedding data. In addition, the IRSS algorithm has outperformed the original one in terms of imperceptibility. This was demonstrated by the achieved PSNR values, which were between 37.28 and 39.74.
- Applying the newly designed and improved algorithms, it has been noticed that the main steganography parameters, reliability, robustness, and imperceptibility have been noticeably improved, as the high PSNR values clearly show. For instance, the range of the obtained values was between 37.28 and 48.30 dB. As for robustness, it has been measured using different types of attacks. Using such attacks did not affect the reliability of the steganography system; accordingly, all hidden information was successfully retrieved with 100% recovery.
- Improving the stego-system robustness by ECC insertion and by adding redundancy bits to the secret embedded message has been defined and evaluated using four different embedding methods. Based on the results, the RS-code and redundancy bits have been exploited to enhance the robustness feature of the CR-BIS and IRSS algorithms respectively.

5.3 Recommendations

A list of research problems that can be investigated as a possible extension to the research work reported in this book is presented below:

1. In this book, all the proposed techniques select grayscale images as the cover-image (Abdul-mahdi et al., 2013). However, color images are widely used by millions of users worldwide over the Internet. Therefore, all the proposed techniques could be further extended for hiding secret information in color images. When a color image is used as the cover-image, the proposed techniques can be directly applied to one of the color planes of the cover-image.
2. The current technology stimulates the indirect deployment of steganography into portable devices such as the iPhone. Hence, the proposed algorithm needs to be revisited in order to improve its functionality with a customized outlook that fits such devices and their bandwidth. To this end, enhancements against severe image compression are necessary.
3. In image steganography scenarios, it is hard to come up with a single scheme that fits all image modalities. Consequently, the solution to this problem is to use artificial intelligence to optimize the performance by: adjusting the characteristic region size; selecting the suitable data-hiding scheme; selecting the proper parameters for compressing the characteristic regions; and depending on the image modality and on the characteristics.
4. The CR-BIS algorithm can be extended to hide the secret information in video files rather than in digital image files to improve the hiding-payload capacity. Consequently, different types of attacks such as cropping, resampling, scanning, rotation, and resizing can be used to assess the hiding algorithm.
5. Further work needs to be done to decide the most appropriate ECC which can give better results in enhancing the steganographic system robustness. Low-density parity-check (LDPC) could be the recommended option, since LDPC codes are increasing used in applications requiring a reliable and highly efficient information transfer.

References

Abdul-mahdi, N. H., Yahya, A., Ahmad, R. B., & Al-Qershi, O. M. (2013). Secured and robust information hiding scheme. *Procedia Engineering Journal, 53*, 463–471.

Hamid, N., Yahya, A., Ahmad, R. B., & Al-Qershi, O. M. (2012a). Image steganography techniques: An overview. *International Journal of Computer Science and Security (IJCSS), 6*(3), 168–178.

Hamid, N., Yahya, A., Ahmad, R. B., & Al-Qershi, O. M. (2012b, April). *Characteristic Region Based Image Steganography Using Speeded-Up Robust Features Technique*. Paper presented at the 1st International Conference on Future Communication Network (ICFCN'12). IEEE international conference, Iraq, Baghdad.

Hamid, N., Yahya, A., Ahmad, R. B., Najim, D., & Kanaan, L. (2013a). Steganography in image files: A survey. *Australian Journal of Basic and Applied Sciences, 7*(1), 35–55.

Hamid, N., Yahya, A., Ahmad, R. B., Najim, D., & Kanaan, L. (2013b). Enhancing the robustness of digital image steganography using ECC and redundancy. *WULFENIA Journal, 20*(4), 153–169.

Kundur, D., & Hatzinakos, D. (1998). *Towards a telltale watermarking technique for tamper-proofing*. Image Processing, 1998. ICIP 98. Proceedings,Chicago, IL, USA.

Mali, S. N., Patil, P. M., & Jalnekar, R. M. (2012). Robust and secured image-adaptive data hiding. *Digital Signal Processing: A Review Journal, 22*(2), 314–323.

Reed, I. S., & Solomon, G. (1960). Polynomial codes over certain finite fields. *Journal of the Society for Industrial and Applied Mathematics, 8*(2), 300–304.

Zhicheng, N., Yun-Qing, S., Ansari, N., & Wei, S. (2006). Reversible data hiding. *IEEE Transactions on Circuits and Systems for Video Technology, 16*(3), 354.

Appendix A
Summary of the Drawbacks of the Existing Steganographic Techniques and the Benefits of the CR-BIS Algorithm

Method	Description
Spatial domain techniques	Large payload capacity but vulnerable even to simple attacks
	Not robust against JPEG compression and image filtering attacks
	Not robust against rotation, cropping, and translation
	Not robust against noise
	Many work only on the BMP format
DCT domain techniques	Less prone to attacks than the former methods at the expense of capacity
	Penetrated by second-order statistics
	Penetrated by DCT coefficients distribution
	Employed only on the JPEG format
	Double-compresses the file
	Not robust against rotation, cropping, and translation
	Not robust against noise
	Not robust against modification of quantization table (i.e., re-compression)
CR-BIS algorithm	Small embedding space at the benefit of robustness; resolved by targeting video files
	Resistance to rotation, translation, cropping, and moderate noise impulses
	No known statistical vulnerabilities
	Resistance to lossy compression due to embedding in DWT domain
	Performs better than DCT algorithms in keeping the carrier distortion to a minimum

A. Yahya, *Steganography Techniques for Digital Images*,
https://doi.org/10.1007/978-3-319-78597-4

Index

A. Yahya, *Steganography Techniques for Digital Images*,
https://doi.org/10.1007/978-3-319-78597-4

MIX
Papier aus verantwortungsvollen Quellen
Paper from responsible sources
FSC® C105338

If you have any concerns about our products,
you can contact us on
ProductSafety@springernature.com

In case Publisher is established outside the EU,
the EU authorized representative is:
Springer Nature Customer Service Center GmbH
Europaplatz 3, 69115 Heidelberg, Germany

Printed by Libri Plureos GmbH
in Hamburg, Germany